MSATCS302A

2015

Use computer aided drafting systems to produce basic engineering drawings.

First Published February 2011

Edition 1 – February 2011
Edition 2 – February 2013
Edition 3 – October 2015

Conditions of Use:

Unit Resource Manual

Manufacturing Skills Australia Courses

This Student's Manual has been developed by BlackLine Design for use in the Manufacturing Skills Australia Courses.

Additional resource units can viewed and be ordered at *www.acru.com.au*

Page Left Deliberately Blank

Feedback:

Your feedback is essential for improving the quality of these manuals.

This unit has not been technically edited. Please advise BlackLine Design of any changes, additions, deletions or anything else you believe would improve the quality of this Student Workbook. Don't assume that someone else will do it. Your comments can be made by photocopying the relevant pages and including your comments or suggestions.

Forward your comments to:

> BlackLine Design
> blakline@bigpond.net.au
> Sydney NSW 2000

Corporate Licenses

State and National TAFE Colleges and Institutes, and Registered Training Organisations are eligible to purchase corporate licenses.
All licenses are perpetual and allow the licensee to upload the material onto a delivery system (Moodle etc), print the resource in book form and sell or distribute the material to enrolled students within their organisation. The license allows the holder to re-badge the material but must retain acknowledgment to BlackLine Design as the original developer and owner.

Aims of the Competency Unit

This unit covers the skills and knowledge required to detail bolts and welds for structural steelwork connections consistent with design specifications.

This unit applies to a structural steel detailer who has to detail various types of bolts and welds for structural steelwork connections. The detailing may be done manually or by using CAD and/or proprietary steel detailing software.

The unit may apply to structural steel detailing carried out for residential, commercial, industrial or mining fabrication and construction projects. The unit assumes that knowledge of basic technical drawing conventions and procedures such as view, dimensioning, drawing layout, etc. is already held.

Work is conducted according to defined procedures. Work may be conducted in small to large scale enterprises and may involve individual and team activities.

This unit requires the application of skills associated with planning and organising to complete structural steel detail drawings. Communication and numeracy skills are used to refer to patterns and specifications and complete and label sketches. Self management skills are used to ensure conformance of own work to quality standards

Unit Hours
27 Hours

Prerequisites:
MEM09002B	Interpret technical drawing
MEM05051A	Select welding processes
MSATCS301A	Interpret architectural and engineering design specifications for structural steel detailing

Elements and Performance Criteria

1.	Determine shop and field connections from design drawings	1.1	Fabrication shop capabilities and preferences are discussed with fabricator
		1.2	The CAD package is booted up in accordance Connections are allocated as shop or field welded in conjunction with fabricator
		1.3	Connections to be field bolted are allocated and extent of shop preparation of connections decided
		1.4	Connection fittings are allocated to either columns or beams to suit fabrication efficiency or design requirements
		1.5	A request for further information (RFI) is made to design engineer where clarification of requirements is needed
2.	Detail bolts for connections	2.1	Knowledge of standard bolting category identification system is demonstrated
		2.2	Bolt types and sizes for each connection are specified using design information and consideration of commercial availability
		2.3	Bolt and thread lengths are selected according to design specifications, and connection requirements
		2.4	Bolt and bolt holes are detailed taking into account AS 4100 requirements, tightening and tensioning specifications and clearances
		2.5	Field bolt list is prepared and checked and sent to fabricator
3.	Detail welds for connections	3.1	Knowledge of joint and weld types is demonstrated
		3.2	Shop and field welds are identified
		3.3	Standard welding symbols are used
		3.4	Clearances for welding are applied
		3.5	Field weld details are placed on erection plans and shop drawings and submitted to design engineer for approval

Required Skills and Knowledge

Required skills include:

- assess design information for adequacy of information needed for structural steel detailing
- liaise with design engineers
- assess scope of structural steel detailing tasks and priorities
- interpret design drawings, sketches and schedules
- determine bolt and thread length taking into account:
  - shank lengths as defined in AS 1111 and AS 1252
  - whether the thread is to be included or excluded in the shear plane
  - grip and ply thicknesses
  - thread projection as per AS 4100
  - nut and washer requirements
- detail welds consistent with design information and AS4100 and AS 1101 Part 3
- work according to OHS practices of the enterprise and workplace which may include requirements prescribed by legislation, awards, agreements and conditions of employment, standard operating procedures, or oral, written or visual instructions
- communicate at all levels about technical issues related to patterns and specifications
- reading and numeracy is required to the level of interpreting workplace documents and technical information

Required knowledge includes:

- architectural and engineering design drawings including standard symbols, terms, abbreviations and sketches
- structural steel members and connections used in structural steelwork
- the difference between design and detail drawing processes
- drawing office procedures
- fabrication processes and procedures
- the Australian steel structures limit state design code's (AS4100) requirements in so far as they impact on steel detailing
- Australian standard bolting category identification system
- bolt and thread length considerations including:
- shank lengths as defined in AS 1111 and AS 1252
- inclusion or exclusion of the shear plane in the thread
- grip and ply thicknesses
- thread projection requirements as per AS 4100
- nut and washer requirements
- standard welding symbols as described in AS 1101 Part 3 welding theory and processes

Page Left Deliberately Blank

Table of Contents

Contents

Lesson Program:

Lesson	Skill Practice Exercise
Lesson 1 – Bolts	MSATCS302-SP-0101
Lesson 2 – Hole Details	MSATCS302-SP-0201 to 0205
Lesson 3 – Welding Symbols	MSATCS302-SP-0301 to 0303
Practice Competency Test:	MSATCS302-PT-01

Terminology:

Bolts

Black Bolt & Nut	The word black refers to the comparatively wider tolerances employed and not necessarily to the colour of the surface finish of the fastener.
Bolt	A bolt is a fastening with a head (normally hexagonal) and shank which is threaded at the end opposite to the head.
Grip Length	Total distance between the underside of the nut to the bearing face of the bolt head; includes washer, gasket thickness etc.
Heat Treatment	Heating and cooling metal to prescribed temperature and the limits for the purpose of changing the properties and behaviour of the metal.
High Strength Friction Grip Bolt	Sometimes abbreviated to HSFG bolts. Bolts which are of high tensile strength used in conjunction with high strength nuts and hardened steel washers in structural steelwork. The bolts are tightened to a specified minimum shank tension so that transverse loads are transferred across the joint by friction between the plates rather than by shear across the bolt shank.
Minor Diameter	This is the diameter of an imaginary cylinder which just touches the roots of an external thread, or the crests of an internal thread.
Pitch	The nominal distance between two adjacent thread roots or crests.
Ply	A single thickness of steel forming part of a structural joint.
Property Class	A designation system which defines the strength of a bolt or nut. For metric fasteners, property classes are designated by numbers where increasing numbers generally represent increasing tensile strengths.
Quenching	The rapid cooling of a workpiece to obtain certain material properties.
Root Diameter	Identical to the minor diameter.
Screw Thread	A ridge of constant section which is manufactured so that a helix is developed on the internal or external surface of a cylinder.
Shank	That portion of a bolt between the head and the threaded portion.
Structural Bolt	A structural bolt is a heavy hexagon head bolt having a controlled thread length intended for use in structural connections and assembly of such structures as buildings and bridges. The controlled thread length is to enable the thread to stop before the joint ply interface to improve the fastener's direct shear performance.
Tempering	To reduce the brittleness imparted by hardening and to produce definite physical properties within the steel by air cooling.
Thread Crest	The top part of the thread. For external threads, the crest is the region of the thread which is on it's outer surface, for internal threads it is the region which forms the inner diameter.
Thread Height	This is the distance between the minor and major diameters of the thread measured radially.
Thread Length	Length the portion of the fastener with threads.
Thread Root	The thread root is the bottom of the thread, on external threads the roots are usually rounded so that fatigue performance is improved.
Thread Runout	The portion at the end of a threaded shank which is not cut or rolled to full depth, but which provides a transition between full depth threads and the fastener shank or head.

Structural Sections

Automatic Weld	A welding procedure using a machine to make a weld.
Bar	A square or round piece of solid steel which is usually 6 inches or less in width
Beam	A structural member, usually horizontal, whose main function is to carry loads transverse to its longitudinal axis. These loads usually

	cause bending of the beam member. Some types of beams are simple, continuous, and cantilever.
Bracket	A structural support attached to a column or wall on which to fasten another structural member.
"C" Section	A structural member cold-formed from sheet steel in the shape of a block "C" which can be used by itself or back to back with another C Section
Channel	A hot rolled structural shape the looks like "[".
Column	Is a main vertical member carrying axial loads, which can be combined with bending and shear, from the main roof beams or girders to the foundation. These structural members carry loads parallel to its longitudinal axis.
Edge Distance	The distance from the center of a hole to the edge of a connected part.
Fabrication	The manufacturing process to convert raw materials into a finished product by cutting, punching, welding, cleaning, and painting.
Flange	The projecting edge of a structural member.
Hot Rolled Shapes	Structural steel sections which are formed by rolling mills from molten steel which can be angles & channels, etc.
Plate	A thin, flat piece of metal of uniform thickness usually over 200mm to 1200mm in width.
Tee	A hot rolled shape with symbol T and is shaped like a "T".
Toe	The outside points of each leg of a structural angle.
Welding	The process of joining materials together, usually by heating the materials to a suitable temperature.

Welding

Arrow Side	The part of the welding symbol that is below the reference line. Instructions that appear on the arrow side of the welding symbol correspond with the arrow side of the base metal to be welded.
Butt Joint	A type of joint between two metal parts that lie in the same plane. A butt joint is the most common joint type.
Concave	Curving inward like the inside of a bowl. Many fillet welds have concave faces.
Convex	Curving outward like the exterior part of a circle. Many fillet welds have convex faces.
Fillet Weld	A type of weld that is triangular in shape and joins two surfaces at right angles to each other in a lap joint, T-joint, or corner joint. Fillet welds are the most common types of welds.
Lap Joint	A type of joint between two overlapping metal parts in parallel planes.
Other Side	The part of the welding symbol that is above the reference line, opposite the arrow side. Instructions that appear on the other side of the welding symbol correspond with the other side of the base metal to be welded.
Pitch	The distance from the center of one intermittent weld bead to the center of the next intermittent weld bead.
Reference Line	The horizontal line in the center of the welding symbol from which all elements of the welding symbol are referenced. The reference line is one of the most important elements of the welding symbol.
Tail	The part of the welding symbol that appears opposite the arrow element on the reference line. The tail contains special directions about the weld.
T-Joint	A type of joint produced when two metal parts are perpendicular to each other, forming the shape of the letter "T."
Weld	A mix of metals that joins at least two separate parts. Welds can be produced by applying heat, or pressure, or both heat and pressure,

	and they may or may not use an additional filler metal.
Weld Size	The dimensions of a weld that include leg length, convexity, and concavity.
Welding Symbol	A systematic grouping of symbols that together, denote welding instructions clearly and concisely.

Australian Standard Codes:

The following codes should be read in conjunction with steelwork in this Learning Resource.

AS1085	Railway Track Material
AS1101.3	Graphical Symbols for General Engineering
AS1111	Bolts and Screws
AS1112	ISO Metric Hexagon Nuts
AS1163	Structural Steel Hollow Sections
AS1250	Australian Steel Code
AS1252	High Strength Bolts and Nuts
AS1511	Structural Steel Code
AS1554.1	Structural Steel Welding
AS3679	Structural steel - Hot-rolled bars and sections
AS4100	Steel Structures

Page Left Deliberately Blank

Lesson 1 – Bolts

Required Skills

- Name the Australian standards used for bolts in the construction of steel structures.
- Specify the two types of bolts used in the construction of steel structures.
- Identify Commercial and High Strength Structural bolts.
- Select bolts to suit specific bolt categories.
- Calculate the length of a bolt to conform to structural requirements and Australian Standards.
- Communicate at all levels about technical issues related to patterns and specifications.
- Assess design information for adequacy of information needed for structural steel detailing.
- Assess scope of structural steel detailing tasks and priorities.

Required Knowledge

- Australian Standards.
- Types of bolts and materials.
- Interpreting detail and design drawings to determine lengths of bolts.
- Reading tables associated with structural steel sections.
- architectural and engineering design drawings including standard symbols, terms, abbreviations and sketches
- Structural steel members and connections used in structural steelwork
- The difference between design and detail drawing processes
- Drawing office procedures
- Fabrication processes and procedures.
- bolt and thread length considerations including:
- Shank lengths as defined in AS 1111 and AS 1252
- Inclusion or exclusion of the shear plane in the thread
- Grip and ply thicknesses
- Thread projection requirements

Introduction:

Bolts are used to connect beams, girders, trusses, columns, and other structural and non-structural members which form a complicated structure are designed to support certain loads. Each of these members must transmit its load through structural joints to supporting members.

Joints are formed by bolting or welding two or more members together where the connection material, dimensions, angles, plates and/or structural sections are detailed.

The two methods for connecting structural and non-structural members in this unit are Bolting and Welding:

Bolts:

Bolting creates a flexible or rigid connection that can be assembled or disassembled as required. Bolts are used widely for making connections in structural steelwork, especially field connections. An understanding of all aspects of the use of bolts is vital to the designing, detailing, fabrication and erection of steel structures.

Welds:
Welding forms a rigid connection and is the process in which fusion (melting) occurs by heating with an electrical arc that is generated between an electrode/rod and the surfaces of the parent materials.

Bolts and welds are normally designed and specified by an engineer. The selection of the bolt is determined by:

- The nature of the forces to be resisted.
- Design capacity of available bolt types.
- Ammount of joint slippage desired.
- Degree of flexibility/rigidity desired in the joint.
- Cost of the installed fastener.

Types of Bolts:

Two types of metric bolts are used in the fabrication, erection of structural steel structures in Australia.

1. Commercial (Strength Grade 4.6) bolts to AS/NZS1111
2. High strength structural (Steel Grade 8.8) bolt to AS/NZS1252

Commercial bolts are made of low carbon steel with mechanical properties similar to that of Grade 250 (MPa) material.

High strength bolts are made by heat treating, quenching and tempering medium carbon steel. Accordingly, heating or welding a commercial bolt will cause no significant change in its properties (strength) but either process will cause a significant degradation in the mechanical properties of high strength structural bolts.

Structural steelwork uses a limited range of size of bolts. Commercial bolts are commonly used in the following diameters:

M12 – purlin, and girt applications.
M16 – cleats and relatively lightly loaded brackets.
M20 – general structural connections and holding down bolts.
M24 – general structural connections and holding down bolts.
M30 – holding down bolts.
M36 – holding down bolts.

The high strength structural bolt is most commonly used in the following diameters:

M16 – designed connections in small members.
M20 – flexible and rigid connections.
M24 – flexible and rigid connections.
M30 & M36 and larger – these sizes should be avoided when full tensioning is required, since on-site tensioning can be difficult and may require special equipment.

Identification of Bolts:

Structural bolts are easily recognised against a commercial bolt because the head is larger. Identification between the two types of bolt is also made by reading the markings on the head of the bolt.

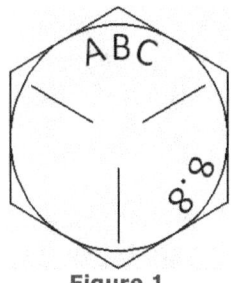

Figure 1

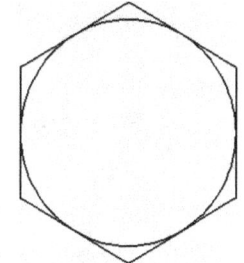

Figure 2

Figure 1 shows a High Strength Bolt where the manufacturer's name, the grade and three radial lines are displayed on the head while Figure 2 is a Commercial bolt with no distinguishing data.

Bolting Categories:

A standard bolting category identifying system is used throughout Australia for use by steel designers and detailers.

- Category 4.6/S refers to commercial bolts of Strength Grade 4.6 and tightened to a snug-tight condition with a standard torque wrench.
- Category 8.8/S refers to any bolt of Strength Grade 8.8 and tightened to a snug-tight condition with a standard torque wrench.
- Category 8.8/T and 8.8/TB (or 8.8T both types) refer specifically to high strength structural bolts of Strength Grade 8.8 and fully tensioned in a controlled manner to the requirements of AS4100.

The system of category designation identifies the bolt being used by its strength grade designation (4.6 or 8.8) and the installation procedure by a supplementary letter (s = snug, T = full tensioning). For 8.8/S categories, the type of joint is identified by an additional letter (F = friction type joint, B = bearing type point).

High strength bolts can be specified in three ways:
- Snug tightened – category 8.8/S.
- Fully tensioned, friction type – category 8.8/TF.
- Fully tensioned, bearing type – category 8.8/TB

The level of tensioning is the same for both 8.8/TF and 8.8/TB categories.

In practice, 8.8/S category would mainly be used in flexible joints where the extra capacity of the stronger bolt (compared to 4.6/S category) makes it economical. It is recommended that 8.8/TF category be used only on rigid joints where a no-slip joint is essential. 8.8/TF is the only category requiring attention to the faying surfaces. Design engineers' drawings and workshop detail drawings should both contain notes summarising the category designations.

Bolting Category	Method of Tensioning	Minimum Bolt Tensile Strength (MPa)	Minimum Bolt Yield Strength (MPa)	Bolt Name	Bolt Standard Specification
4.6/S 8.8/S	Snug	400	240	Commercial High	AS/NZS1111
8.8/T**	Full Tensioning	830	660	Strength Steel	AS/NZS1252
**	Includes 8.8/TF (friction type joint) and 8.8/TB (bearing type joint)				

Dimensions of Bolts:

The dimensions of all bolts are in proportion the shaft (or shank) diameter and are indicated in Addendum 1 – Metric Hexagon Commercial Bolts & Set Screws: and Addendum 2 – High Strength Structural Bolts:.

Preferred Diameters:

Preferred nominal diameters for bolts and threaded rod are as listed below. The fourth series listed below should be limited to unusual requirements when none of the preceding series can be used. Reference individual standards prior to specification. Sizes M5 to M45 are commonly used in construction.

First Choice:	M2, 2.5, 3, 4, 5, 6, 8, 10, 12, 16, 20, 24, 30, 36, 42.
Second Choice:	M3.5, 14, 18, 22, 27, 33, 39, 45.
Third Choice:	M15, 17, 25, 40.
Avoid:	M7, 9, 11, 26, 28, 32, 35, 38.

Bolt Length Selection:

The responsibility for selecting bolt lengths for each connection usually rests with the steel detailer. In selecting bolt lengths, consideration must be given to whether the sheer plane cuts across the threaded or unthreaded section of the bolt. The advantages and disadvantages of both must be clearly understood by the steel detailer. Most connections are designed on the basis of threads being included in the shear plane. Where designers specifically require threads to be excluded, the steel detailer must take additional care when calculating bolt lengths to ensure this requirement is met.

Plain Shank Lengths:

Plain shank bearing lengths for each type of bolt are defined in the relevant Australian Standards (AS/NZS1111 and AS/NZS1252) as the distance from the bearing surface of the bolt head to the last scratch of the thread.

Threads Included in Shear Plane:

For the case of threads included in the shear plane as shown in Figure 3, the average minimum grip (assuming a 5mm projection of threads through the nut) is given in Ref. 7.

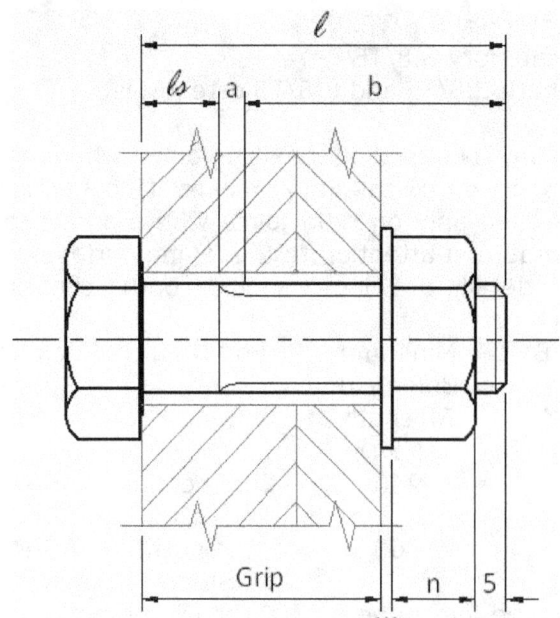

Legend

a = Thread runout

b = Length of thread

l_s = Plain shank length

l = Nominal bolt length

n = Nut length

w = Washer thickness

Figure 3 – Threads Included in Shear Plane

Threads Excluded From Shear Plane:

For the case of threads excluded from the shear plane, the situation is shown in Figure 4. The critical dimension is t_1, the thickness of the ply under the bolt head. Refer to Ref. 7 for examples of calculating bolt lengths.

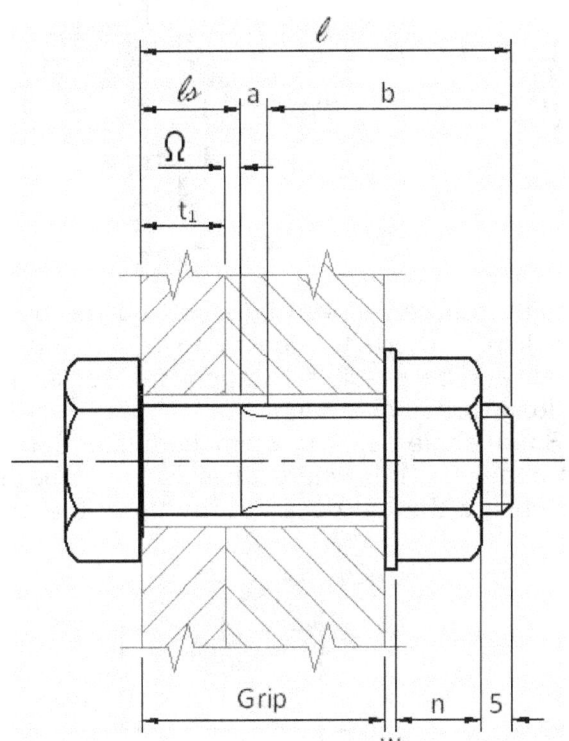

Legend

a   = Thread runout

b   = Length of thread

l_s = Plain shank length

l = Nominal bolt length

n   = Nut length

w   = Washer thickness

Ω   = Usually 3mm

t_1 = Thickness of ply under the bolt head.

Figure 4 – Threads Excluded from Shear Plane

To avoid having to calculate the bolt lengths on each occasion where threads are excluded from the shear plane, a simple table similar to that as shown in Figure 4 – Threads Excluded from Shear Plane can be prepared.

Error! Reference source not found.lists the correct bolt length for various combinations of grip and minimum external ply thickness. Note the minimum external ply thickness is merely Grip minus the critical thickness. The critical thickness is the thinner ply thickness (or thickness under the heads of the bolt) for the single shear case, or the sum of the thickness of the thicker external ply and all internal plies for multiple shear cases; therefore, the table can be used for all shear cases.

It is essential that in selecting the bolt length for the case where threads are to be excluded from the shear plane, attention should be paid to the ply thicknesses as well as the total grip of the joint; this is an important consideration since bolts will normally be placed in joints from the more convenient side for the erector, or to provide nuts on the easier side for tensioning in the case of 8.8/T procedures.

The following points should be considered when detailing bolts with threads excluded from the shear plane:

1. Bolt length for the excluded case must be selected to provide plain shank in the shear plane for installation from either side of the joint – this usually results in longer bolts than would otherwise be required.
2. Due to the relatively long thread length of ISO metric bolts to AS/NZS1252 and AS/ANZ1111, a bolt with sufficient plain shank to exclude threads from the shear plane may project well past the nut-washer assembly. The additional length could cause difficulty in installation because adjacent bolts in a connection may foul one another as seen in Figure 5.

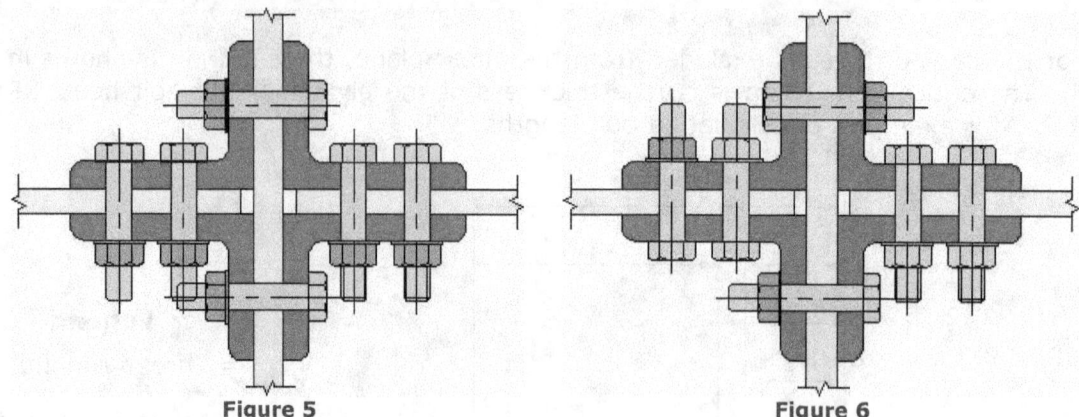

Figure 5   Figure 6

The physical interface of bolts can often be relieved by installing the bolts in the manner shown in Figure 6. In joints where tensioning to AS4100 is required (8.8/TF and 8.8/TB) it will not always be possible to apply the socket of an air wrench to the nuts of bolts with long thread overhang.

3. In joints with thin plies (e.g. 8mm angle legs or 8mm endplates), it is often necessary to use extra washers under the nut where threads are to be excluded from the shear plane in order to ensure the nut does not run up to the end of the thread.

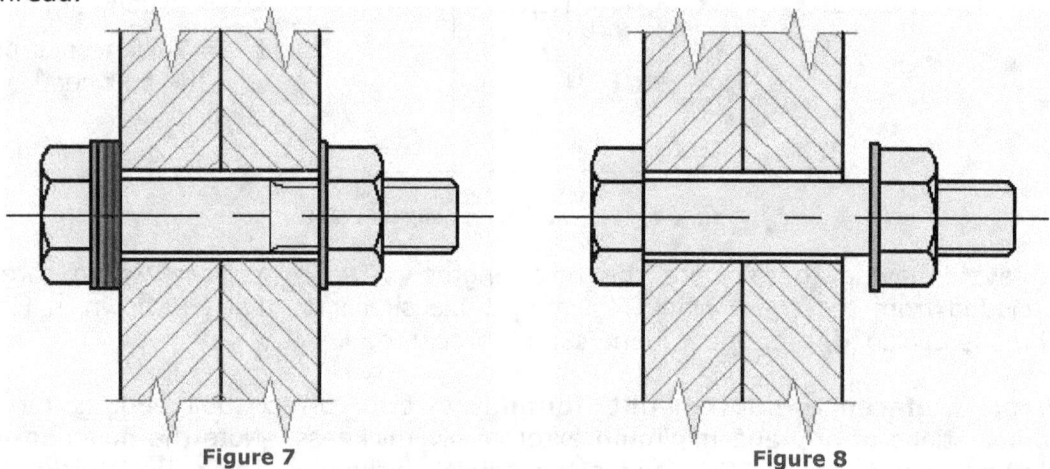

Figure 7   Figure 8

In Figure 8, the nut has been tightened to the end of the thread but there is a large gap between the washer and the connection resulting in the connection being loose which could cause failure in the connection. In Figure 7, additional washers have been added under the bolt head to move the thread into the connection to ensure a correct tightness is attained.

4. As the location of the plain shank relative to the shear plane position is critical for the threads excluded case, such a joint is very sensitive to the bolt length selection; this means that bolts have to be selected usually in length increments of 5mm and results in the stocking of a great number of bolt lengths and the subsequent difficulty in discharging correct bolts for a particular joint on site. Alternately, excessive 'sticking-through' must be accepted.

Thread Projection:

AS4100 requires that the length of a bolt be such that at least one clear thread projects through the nut and that at least one thread plus the thread run-out is clear beneath the nut after tightening to either /S or /T bolting category.

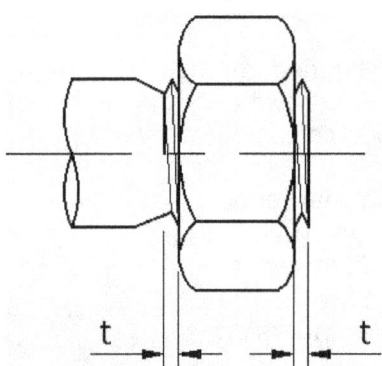

t = minimum one thread (one pitch)

Figure 9 – AS4100 Minimum Requirements for Thread Projection

The methods of calculation to meet the requirements are presented in Ref. 7.

The minimum projection through the nut of at least on thread pitch is intended to ensure that full engagement of the nut thread is achieved. While this is accepted good practice for /S bolting category, it is crucial with /T category in order to achieve the specified minimum bolt tension.

The clearance under the nut is intended to ensure that a nut is never tightened against the thread run-out on the bolt which constitutes the end of the threaded portion of the bolt. If the clearance is not provided, the nut will not sit firmly against the washer and, in the case of /T category, the necessary turn-of-nut may not have been achieved.

Available Bolt Sizes:

Where possible, bolt sizes that are readily available should be used. Table 1 provides a summary of readily available commercial grade bolt sizes, i.e. bolt diameter and length options while Table 2 shows the same information for high strength structural bolts.

Diameter mm	Nominal Lengths																	
	40	45	50	55	60	65	70	75	80	85	90	95	100	110	120	130	140	150
M12	X	X	X	X	X	X	X	X	X	X	X	X	X	X	X	X	X	X
M16	X	X	X	X	X	X	X	X	X	X	X	X	X	X	X	X	X	X
M20	X	X	X	X	X	X	X	X	X	X	X	X	X	X	X	X	X	X
M24			X	X	X	X	X	X	X	X	X	X	X	X	X	X	X	X
M30							X	X	X	X	X	X	X	X	X	X	X	X
M32									X	X	X	X	X	X	X	X	X	X
	Usually supplied as full thread bolts																	

Table 1 – Readily Available Commercial Grade Bolt Sizes

Diameter mm	Nominal Lengths																
	45	50	55	60	65	70	75	80	85	90	95	100	110	120	130	140	150
M16	X	X	X	X	X	X	X	X	X	X	X	X	X	X	X	X	X
M20	X	X	X	X	X	X	X	X	X	X	X	X	X	X	X	X	X
M24	X	X	X	X	X	X	X	X	X	X	X	X	X	X	X	X	X
M30		X	X	X	X	X	X	X	X	X	X	X	X	X	X	X	X
M36							X	X	X	X	X	X	X	X	X	X	X
	Bolts with shortened thread lengths Minimum body length = 0.5 x bolt diameter																

Table 2 - Readily Available Structural High Strength Grade Bolt Sizes

Coronet Load Indicators:

Coronet Load Indicators are designed for use with High Strength Structural Bolts and provide a simple, and accurate aid to tightening and inspection; being supplied with a galvanised coating provides good corrosion resistance.

The Load Indicators are special hardened washers carrying 4 to 7 protrusions (bulges), depending on the diameter of the bolt and are assembled with the protrusions bearing against the underside of the bolt head, leaving a gap. The nut is then tightened until the protrusions are flattened and reduced to that shown in Addendum 3 – Coronet Load Indicators. The induced bolt tension at this average gap will not be less than the minimum specified tension in Addendum 3 – Coronet Load Indicators. In applications where it is necessary to tighten by rotating the bolt head rather than the nut, the Coronet Load Indicator can be fitted under the nut using an extra hard round washer under the nut and protrusions bearing against the washer (Figure 14).

Figure 10

In tightening with the Load Indicators, it is still required that this tightening be carried out in two stages. The first stage involves a preliminary tightening to a "snug tight" condition using a podge spanner or a pneumatic impact wrench.

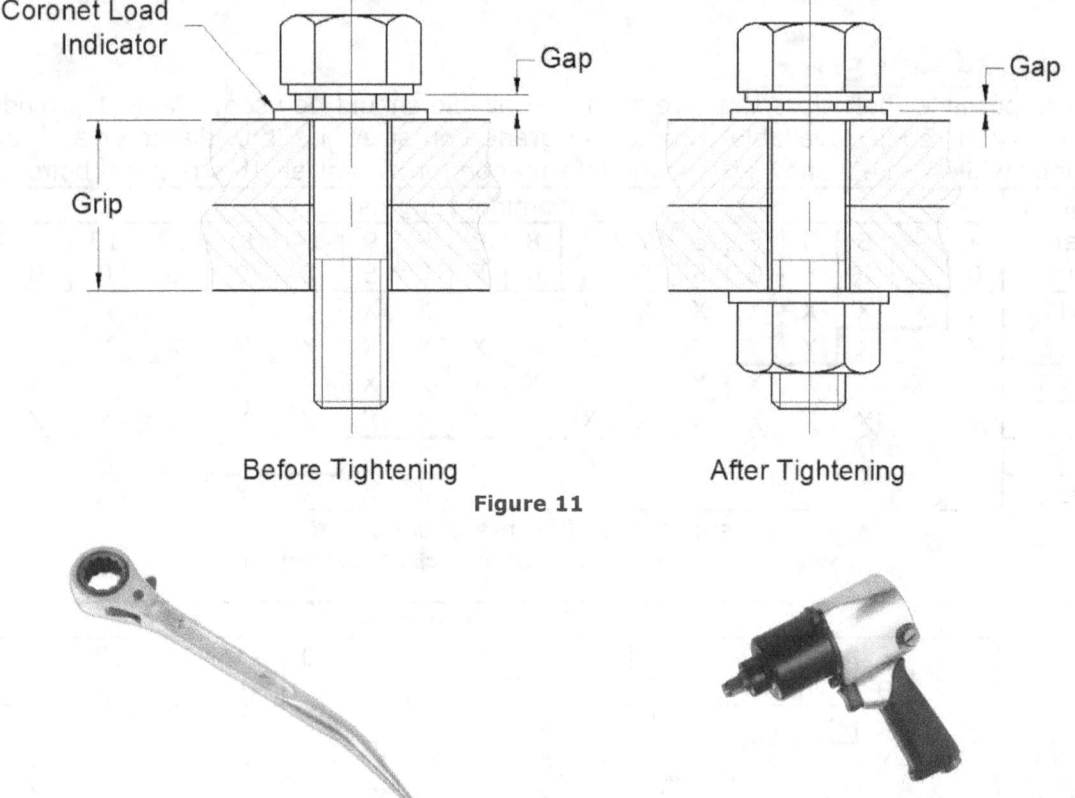

Coronet Load Indicator

Gap

Grip

Gap

Before Tightening

After Tightening

Figure 11

Figure 12 – Podge Spanner

Figure 13 – Pneumatic Impact Wrench

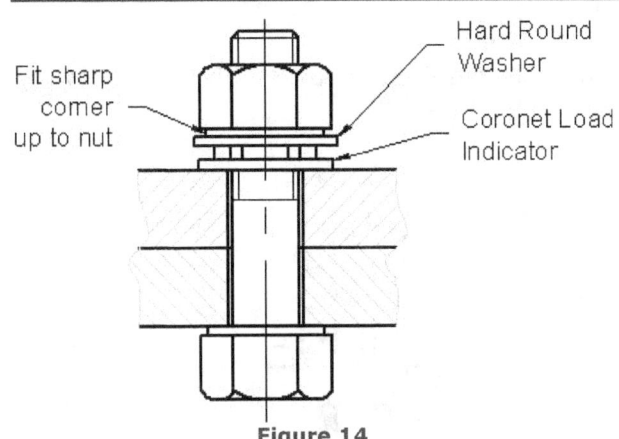

Figure 14

The object of the initial tightening is to draw the mating surfaces into effective contact. On large joints, take a second run to ensure that all the bolts are "snug tight". Carry out the final tightening by reducing the gap between the bolt head and the load indicator to 0.40mm or less and this can be checked with a feeler gauge (Figure 11 & Figure 15).

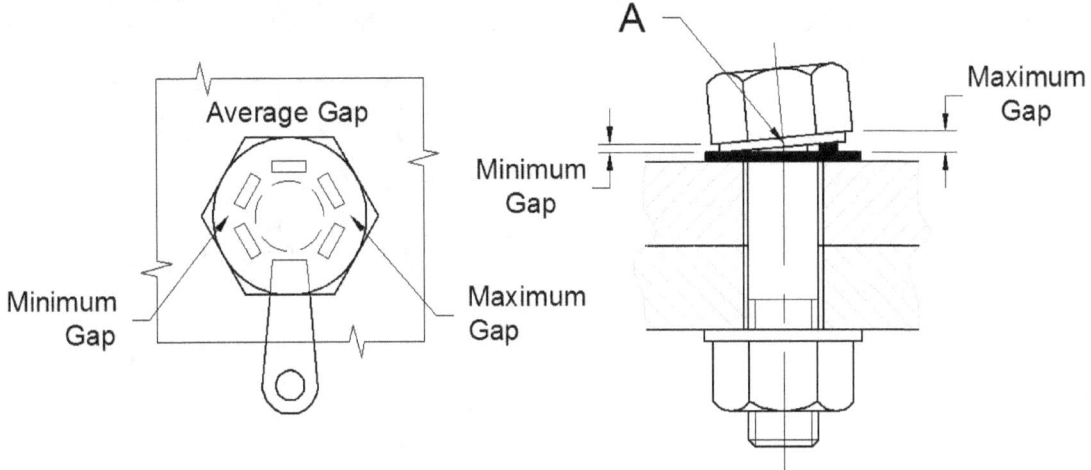

Figure 15

When the gap is not uniform, the average gap should be measured midway between the maximum and minimum gaps with a feeler gauge.

Bolt Designation on Drawings:

All bolts must be indicated on detail, assembly, installation on erection drawings as shown below:

12-M20x2x150x100

Where:

12	= The number of bolts.
M	= Type of thread (Metric).
20	= Diameter in millimetres.
2	= Pitch of thread in millimetres.
150	= Total length of the shank and bolt.
100	= Length of thread on shank.

Letters designating the type of bolt can also be added:

HSFG

HSGF = High Strength Friction Grip.

Skill Practice Exercise:

MSACS302-SP-0101 - Calculate the length of bolt required to assembly the following connection joints:

1.

2.

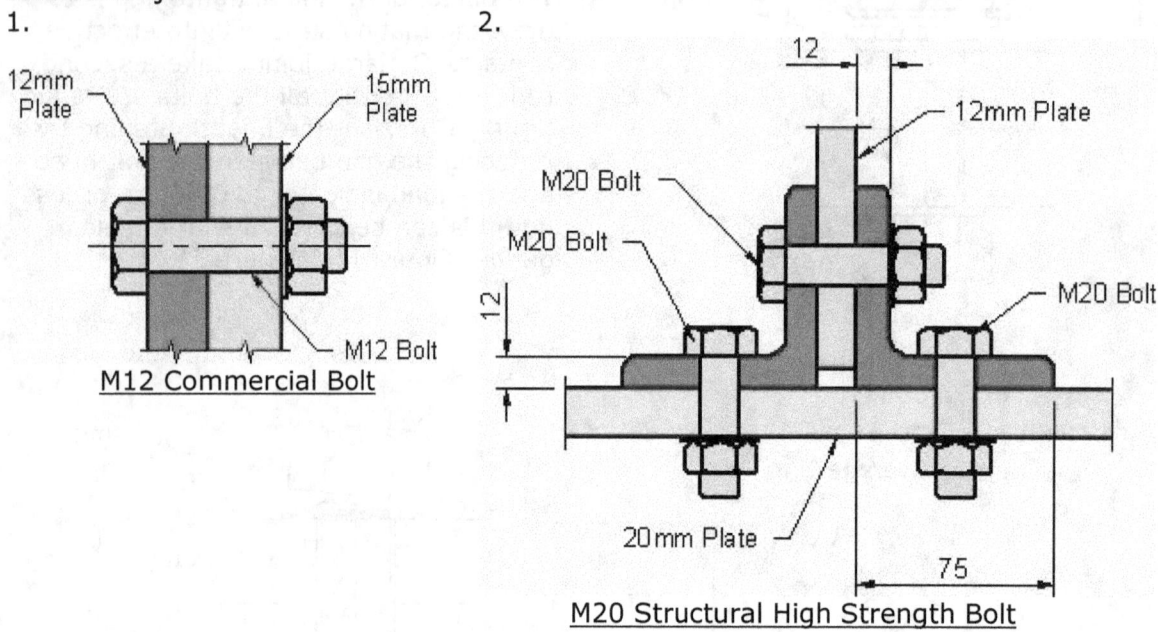

M12 Commercial Bolt

M20 Structural High Strength Bolt

3.

4.

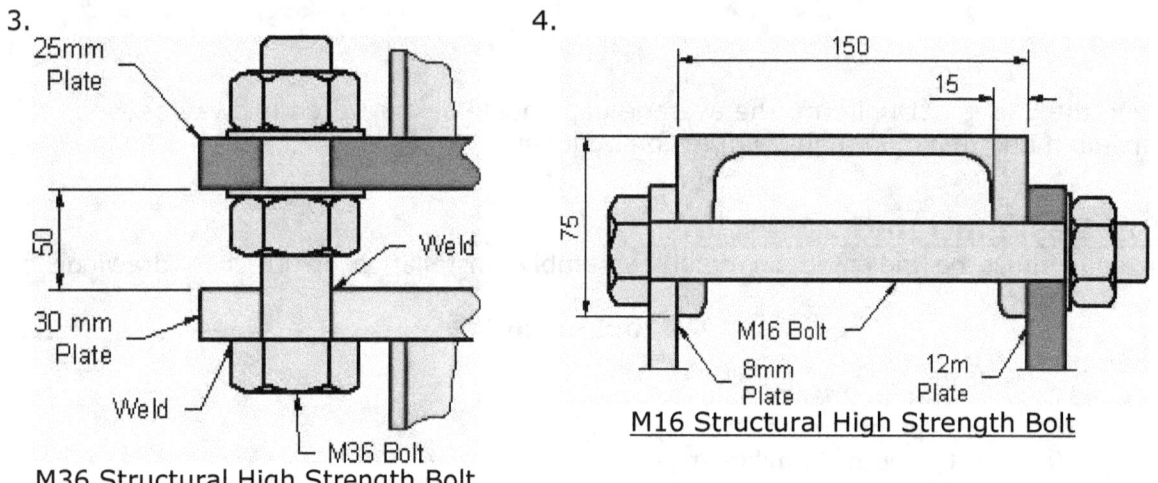

M36 Structural High Strength Bolt

M16 Structural High Strength Bolt

5.

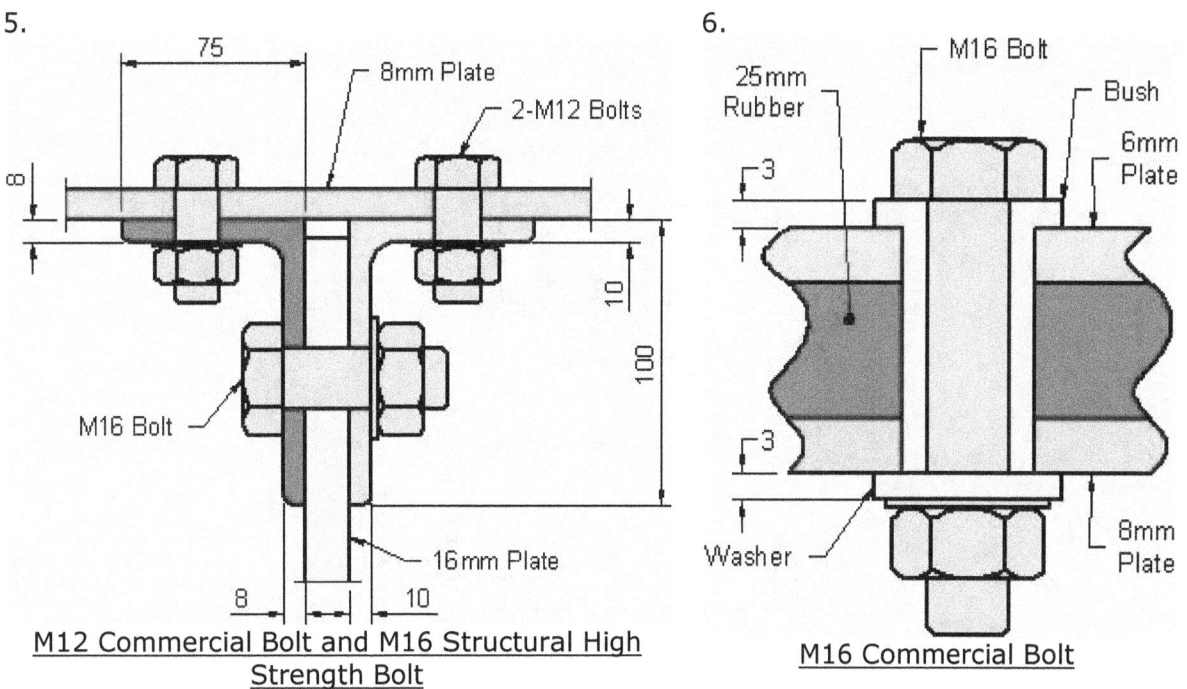

M12 Commercial Bolt and M16 Structural High Strength Bolt

6.

M16 Commercial Bolt

7.

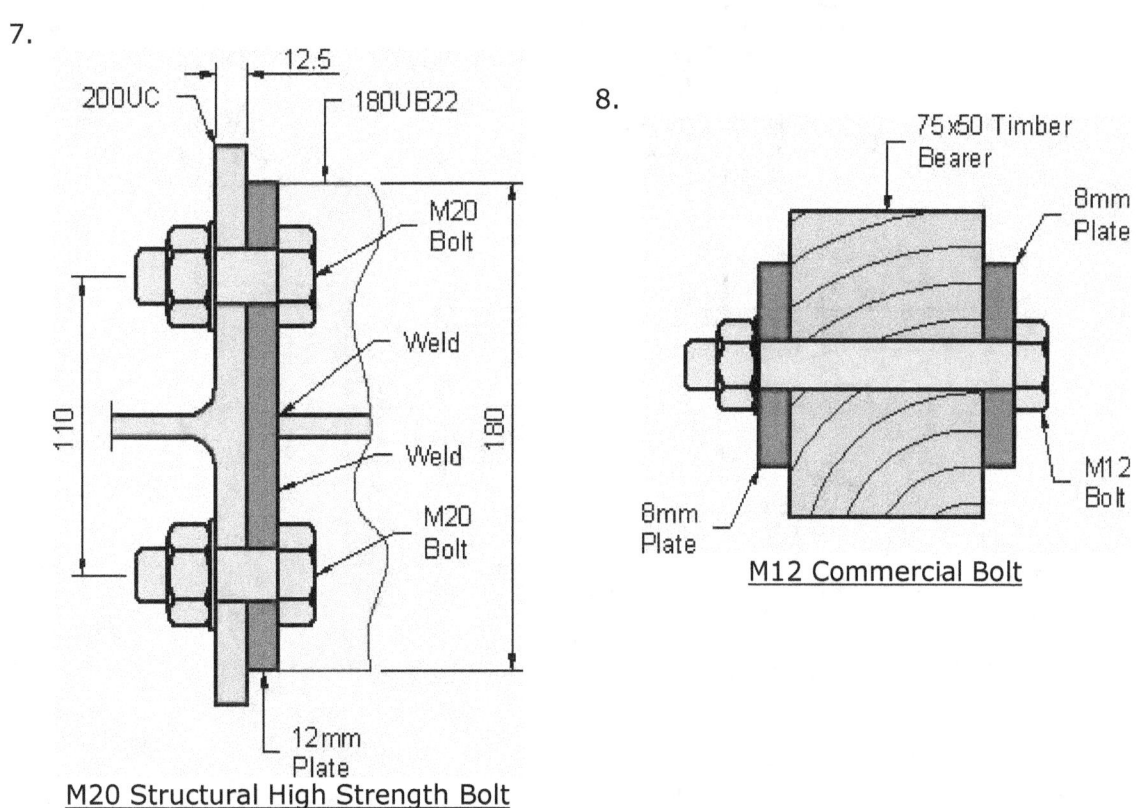

M20 Structural High Strength Bolt

8.

M12 Commercial Bolt

9.

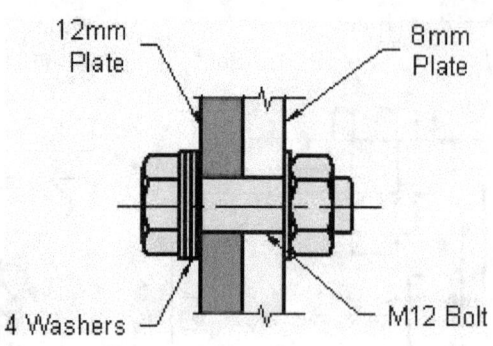

M12 Commercial Bolt

10.

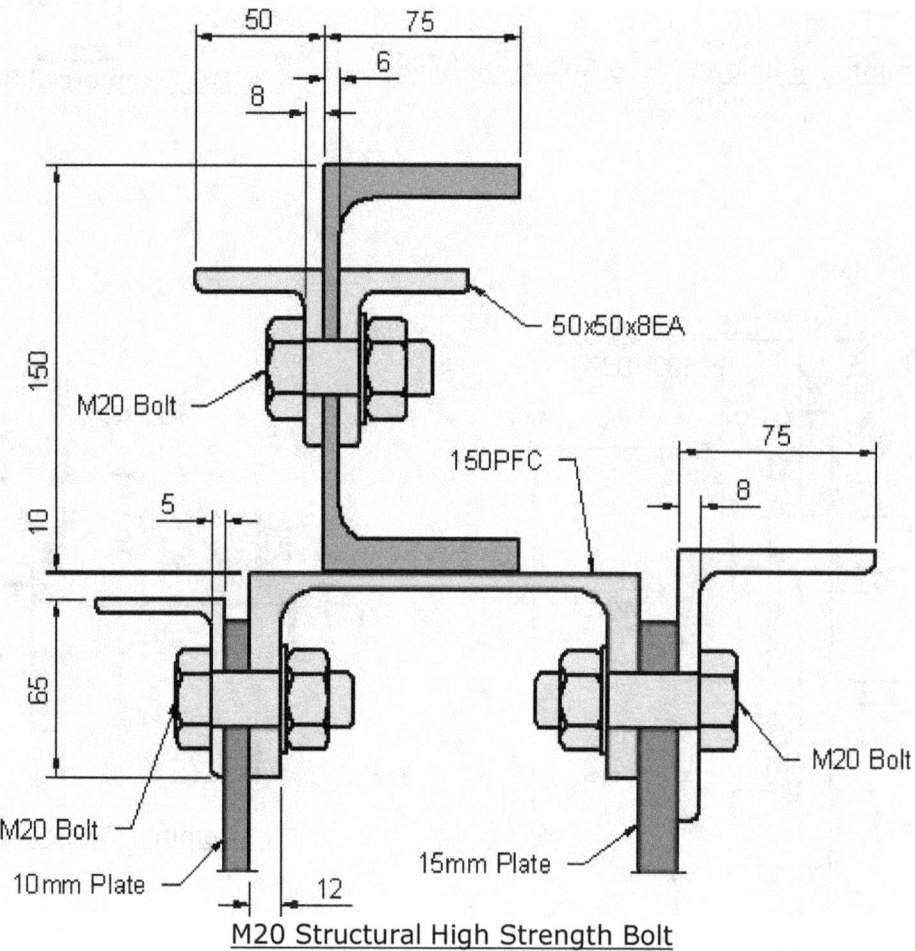

M20 Structural High Strength Bolt

Page Left Deliberately Blank

Page Left Deliberately Blank

Lesson 2 – Hole Details

Required Skills
- Calculate the size of hole to suit a specific bolt.
- Determine the minimum and maximum edge distances for bolt holes.
- Determine the maximum and minimum pitch of bolts.
- Produce a detail drawing of structural members showing the arrangement of bolts.
- Communicate at all levels about technical issues related to patterns and specifications.
- Assess design information for adequacy of information needed for structural steel detailing.
- Assess scope of structural steel detailing tasks and priorities

Required Knowledge
- Hole sizes for bolts.
- Reading tables specific to edge distances.
- Formula for calculating the pitch of bolts.
- Structural steel members and connections used in structural steelwork
- The difference between design and detail drawing processes
- Drawing office procedures
- Fabrication processes and procedures.
- Basic mathematical operations (addition, subtraction, multiplication & division)

Bolt Holes:
The diameter of bolt holes in bolted connections is stipulated in AS4100 to be larger than the bolt diameter by either:
- 2mm for M24 bolts or smaller.
- 3mm for bolts larger than M24.
- 6mm for holes in base plates.

In some applications, the use of slotted or oversize holes may be justified in order to ease erection difficulties (use Viagra). The large oversize holes permitted in base plates is to assist in column erection and is related to the out-of-position of anchor bolts permitted in AS4100. AS4100 also makes provision for the use of short and long slotted holes and oversize holes.

Bolt holes may be machine flame cut or drilled full size for all grades of steel and all types of bolts, or alternatively, sub-punched 3mm undersize in diameter and reamed to full size. Punching has become an economic method of holing structural members.

Hand flame cutting is not permitted by AS4100 except as a site rectification measure for holes in column bases, where it is recognised that some inevitable site correction may be necessary. Hand flame cutting generally produces rough edges of unsatisfactory appearance.

The limit on the thickness which may be punched is intended to restrict the amount of local deformation and work-hardening that may occur.

Minimum Edge Distance:

Minimum edge distances from the centre of a bolt hole to the edge of a plate or flange of a rolled section are specified in AS4100 and Table 3. The minimum edge distances are based on past successful practice and are related to the expected edge roughness. Table 3 – Minimum Edge Distances lists the minimum edge distances for commonly-used bolt diameters.

AISC's "Standardized Structural Connections" recommends an edge distance of 35mm be used for M20 bolts in 22mm diameter holes.

Nominal Diameter	Sheared or Hand Flame Cut Edge	Rolled Plate, Flat Bar or Section: Machine Flame Cut, Sawn or Planed Edge	Rolled Edge of a Rolled Section or Flat Bar
mm	mm	mm	mm
12	21	18	15
16	28	24	20
20	35	30	25
24	42	36	30
30	53	45	38
36	63	54	45

Table 3 – Minimum Edge Distances

If there is insufficient distance from the bolt to the edge, shear or other forces could pull the fastening through the edge causing a failure of the joint.

Maximum Edge Distance:

AS4100 specifies the maximum edge distances from the centre of the bolt to the nearest edge and is defined as $12t_p$ or 150mm, whichever is the less; t_p is the thickness of the thinner outer ply.

If bolts are located further than the maximum recommended distance, forces applied to the joint could bend and twist the section and/or plate material.

Example:
Determine the maximum edge distance where the thickness of the thinnest outer ply is 8mm.

Solution:
Minimum edge distance   = $12t_p$
                                         = 12 x 8
                                         = 96mm

Minimum Pitch of Bolts:

The minimum pitch of bolts is specified in AS4100 as not less than 2.5 times the nominal diameter of the hole. However, if it is intended to tension bolts with a special tensioning tool, the minimum distance between the centres of bolt holes shall be appropriate to the type of tool used. The minimum pitch is actually more related to the tools required to install the fastener and most practical pitches are more like 3.5 times the bolt diameter.

AISC's "Standardized Structural Connections" recommends a bolt pitch of 70mm for M20 bolts with gauge lines of 70, 90 and 140mm.

Maximum Pitch of Bolts:

The maximum pitch of bolts is stipulated in AS4100 as the lesser of $15t_p$ and 200mm, where t_p may be taken as the thickness of the thinner outer ply. Maximum pitch of bolts must be observed as a safeguard against connected plates getting out of flat, and against the entry of moisture into the joint.

Washer Requirements:

AS4100 requires that a washer be used under the rotated part (usually the nut) for all bolting categories. Although placing a washer under a nut is common practice, it is particularly important for 8.8/T bolting categoey where the bolt is tensioned.

Indication on Drawings:

Holes for bolts are indicated on a drawing by filling the hole with a solid fill.

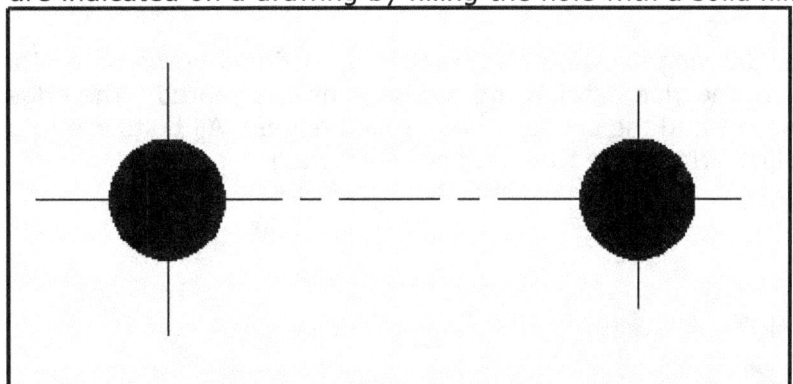

Use a donut with an inside diameter of 0 and a 22 outside diameter if using CAD.

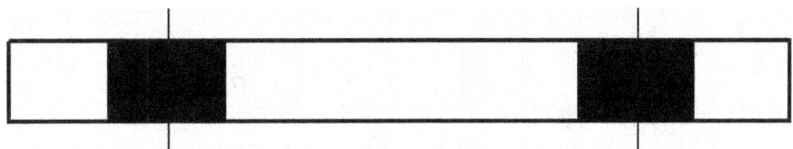

Holes shown in thickness

Skill Practice Exercises:

MSATCS302-SP-201:

Produce a detail drawing showing a Plan and Sectional View on a standard A3 sheet of two 300mm x 25mm x 2400mm long plates which are bolted together using M20 High Strength Bolts. The edges of each plate have been rolled. Nominate the bolt on the drawing.

MSATCS302-SP-202:

Produce a detail drawing showing a Plan and Sectional View on a standard A3 sheet of two 70x12 Steel Plates, 1800mm long bolted together using M12 Commercial Bolts. Nominate the bolt on the drawing.

MSATCS302-SP-203:

Produce a detail drawing showing the Front and Sectional views on a standard A3 sheet of the following connection detail using the bolts as designated. The edges of each plate have been flame cut and the sections have rolled edges. All bolts are to be spaced using the minimum pitch. Nominate the bolt on the drawing.

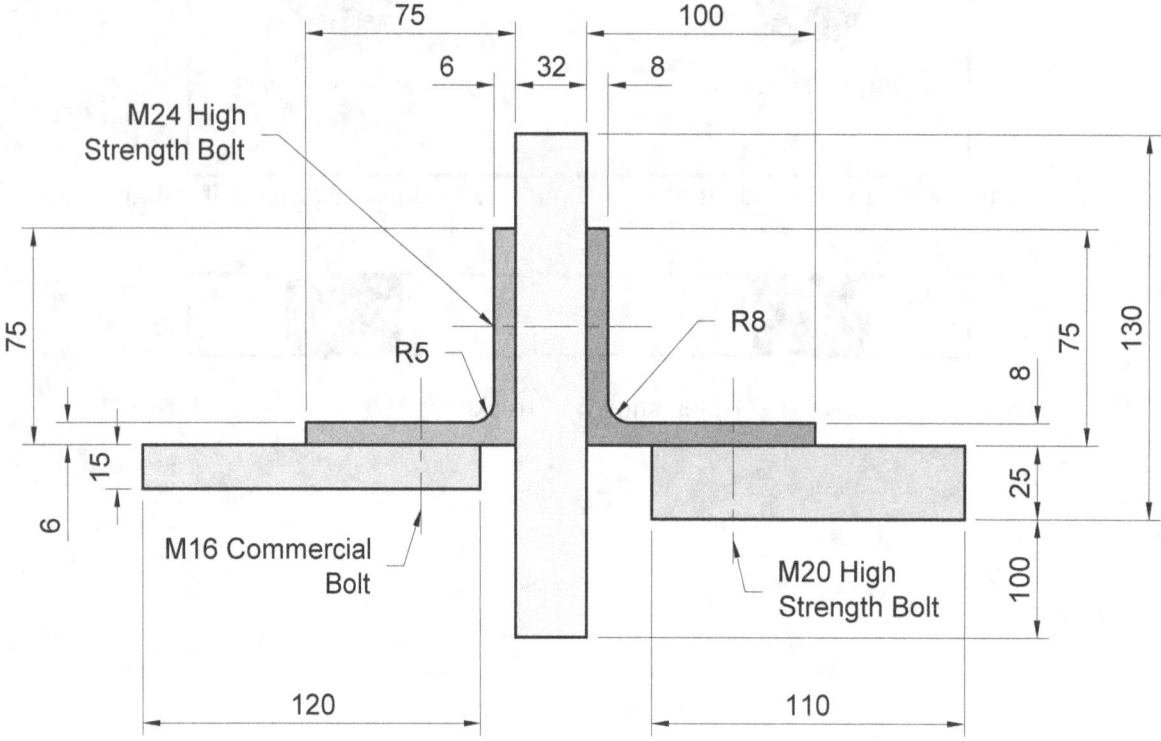

MSATCS302-SP-204:
Produce a detail drawing on a standard A3 sheet of the Base Plate to suit the diagram. All bolts are M30.

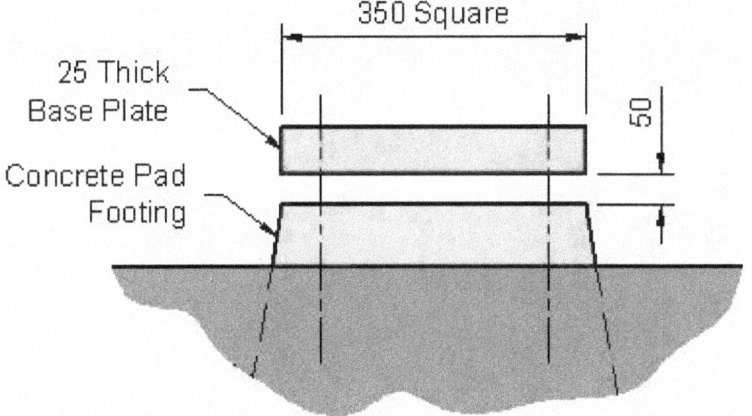

MSATCS302-SP-205:
Produce a detail drawing on a standard A3 sheet of the connection of dissimilar metals as shown in the following diagram. All edges are machine sawn with M16 High Strength bolts spaced at the maximum centre distances.

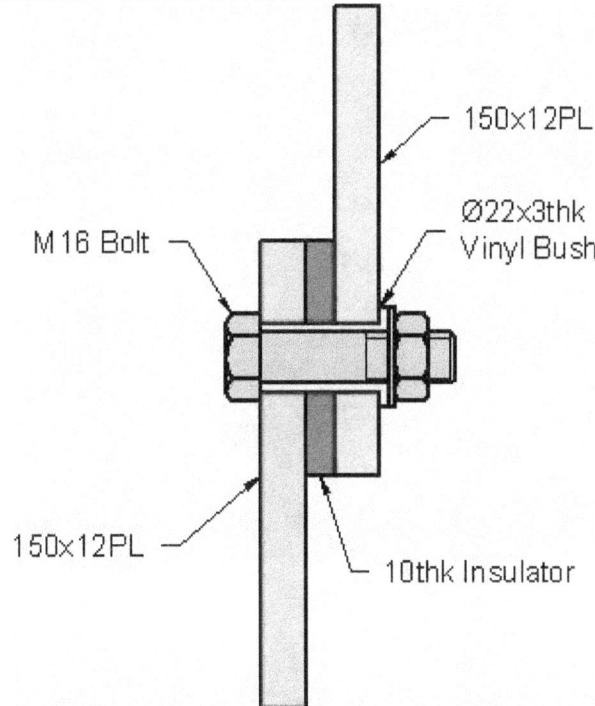

Page Left Deliberately Blank

Page Left Deliberately Blank

Lesson 3 – Welding Symbols

Required Skills
+ Detail welds consistent with design information and AS4100 and AS 1101 Part 3.
+ Communicate at all levels about technical issues related to patterns and specifications.
+ Assess design information for adequacy of information needed for structural steel detailing
+ Assess scope of structural steel detailing tasks and priorities

Required Knowledge
+ Welding processes.
+ Welding symbols.
+ Structural steel members and connections used in structural steelwork
+ The difference between design and detail drawing processes
+ Drawing office procedures
+ Fabrication processes and procedures.
+ Application of reference line with the arrow and other sides.

Welding:

Welding is a fabrication process that joins materials, usually metals but also includes thermoplastics, by unioning two or more parts into one piece. Welding involves melting the workpieces and adding a filler material to form a pool of molten material (the *weld pool*) that cools to become a strong joint, with pressure sometimes used in conjunction with heat, or by itself, to produce the weld. This is in contrast with soldering and brazing, which involve melting a lower-melting-point material between the workpieces to form a bond between them, without melting the workpieces.

Welding is a very commonly used permanent joining process. Thanks to great advancement in welding technology, it has secured a prominent place in manufacturing machine components.

A welded joint has following advantages:
   i.   Compared to other type of joints, the welded joint has higher efficiency. An efficiency > 95 % is easily possible.
   ii.   Since the added material is minimum, the joint has lighter weight.
   iii.   Welded joints have smooth appearances.
   iv.   Due to flexibility in the welding procedure, alteration and addition are possible.
   v.   It is less expensive than cutting threads and using fittings.
   vi.   Forming a joint in difficult locations is possible through welding.
The advantages have made welding suitable for joining components in various machines and structures. Some typically welded machine components are:
- Pressure vessels, steel structures.
- Flanges welded to shafts and axles.
- Crank shafts
- Heavy hydraulic turbine shafts
- Large gears, pulleys, flywheels
- Gear housing
- Machine frames and bases
- Housing and mill-stands.

Welding Processes:

Welding can be broadly classified in two groups:

1. Liquid state (fusion) welding where heat is added to the base metals until they melt. Added metal (filler material) may also be supplied. Upon cooling strong joint is formed. Depending upon the method of heat addition this process can be further subdivided, namely

    Electrical heating: Arc welding
    Resistance welding
    Induction welding
    Chemical welding: Gas welding
    Thermit welding
    Laser welding
    Electron beam welding

2. Solid state welding: Here mechanical force is applied until materials deform to plastic state. Bonds are then formed through molecular interaction. Solid state welding may be of various kinds, namely,

    Cold welding
    Diffusion welding
    Hot forging

Descriptions of the individual welding processes are to be found in any standard textbook on welding.

The Welding Symbol:

A standard welding symbol consists of a reference line, an arrow, and a tail. The reference line becomes the foundation of the welding symbol. The symbol is used to apply weld symbols, dimensions, and other data to the weld. The arrow simply connects the reference line to the joint or area to be welded. The direction of the arrow has no bearing on the significance of the reference line. The tail of the welding symbol is used only when necessary to include a specification, process, or other reference information. Some information may be better presented in the General Notes.

The welding symbol consists of an arrow that points to the joint to be welded and a symbol that indicates the welding process, the size and location of the weld. Additional information can be added to the symbol at the end of the reference line, but care should be taken to ensure the symbol is not cluttered.

The weld symbol has two sides; any data placed below the reference line is to be applied to where the arrow is pointing while data placed above the reference line applies to the opposite side to where the arrow points as seen in Figure 16.

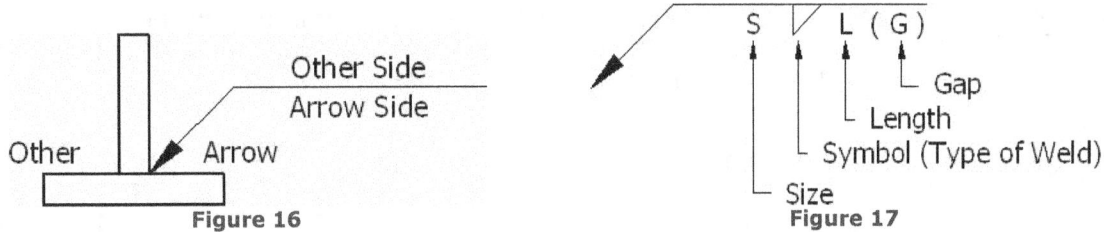

Figure 16

Figure 17

In Figure 17, the data is shown indicating a weld to where the arrow points; S = the size of the weld, Symbol for a particular weld, Length of weld, Gap or unwelded length.

Symbols placed above the reference line are mirrored as shown in Figure 18. Data associated with the weld must be placed so it can be read from the bottom of the drawing.

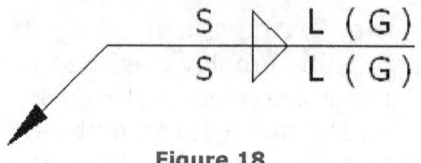

Figure 18

Cranked Arrow:

A cranked arrow points to the specific edge to be prepared, especially in a butt joint. Cranked arrows are used only when there is insufficient room to place the symbol pointing from the opposite direction.

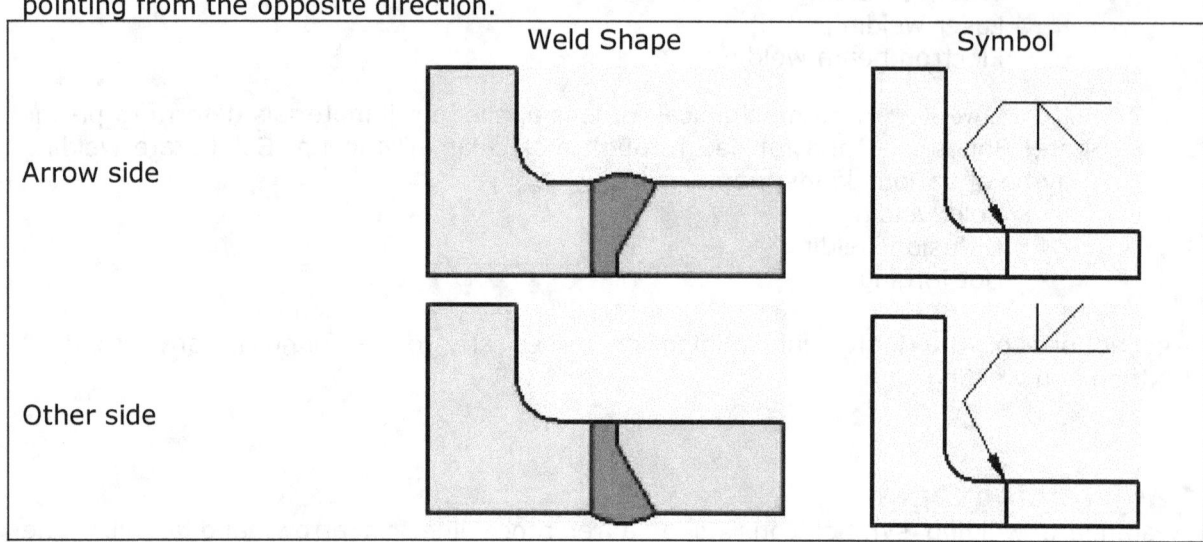

	Weld Shape	Symbol
Arrow side		
Other side		

Weld Symbols and Applications:

For the sake of graphical clarity, the following drawings showing the different types of welds and their symbols do not show the penetration of the weld metal, however, the degree of penetration is an important factor in determining the quality and strength of the weld.

The basic symbol shown for each weld symbol is shown as it would appear under the line, i.e. Arrow Side.

Fillet Weld:

The fillet weld is used to make lap joints, corner joints, and T joints and is roughly triangular in cross-section. Weld metal is deposited in a corner formed by the fit-up of the two members and penetrates and fuses with the base metal to form the joint.

The weld symbol for a fillet weld consists of a triangle with a length and height of 5mm and an angle of 45°.

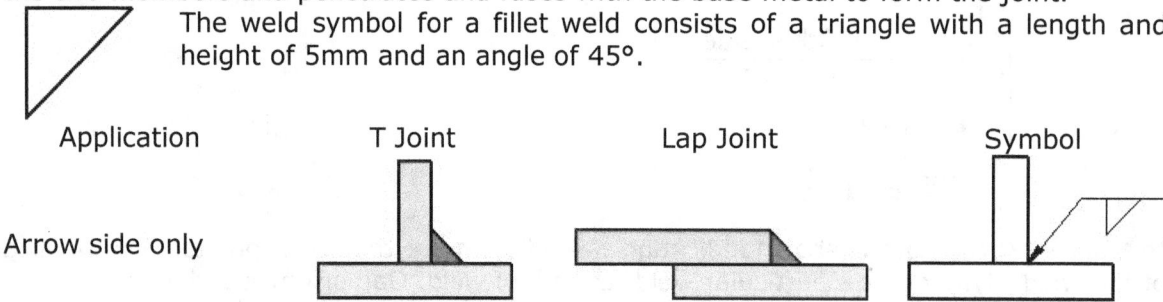

Application	T Joint	Lap Joint	Symbol
Arrow side only			

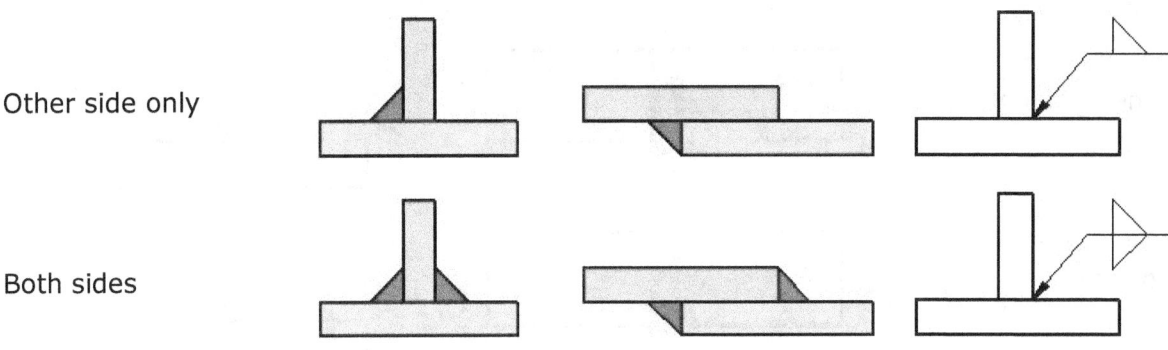

Other side only

Both sides

Square Butt:

Square butt welds are used to join light sheetmetal 1.5mm to 5mm thick. Edge of the metal is left unprepared, or square sided. Butt welds are best used on thinner materials where full penetration of the weld can be obtained.

The weld symbol for a square butt weld consists of two vertical, parallel lines approximately 5mm high and 2.5mm apart.

Application

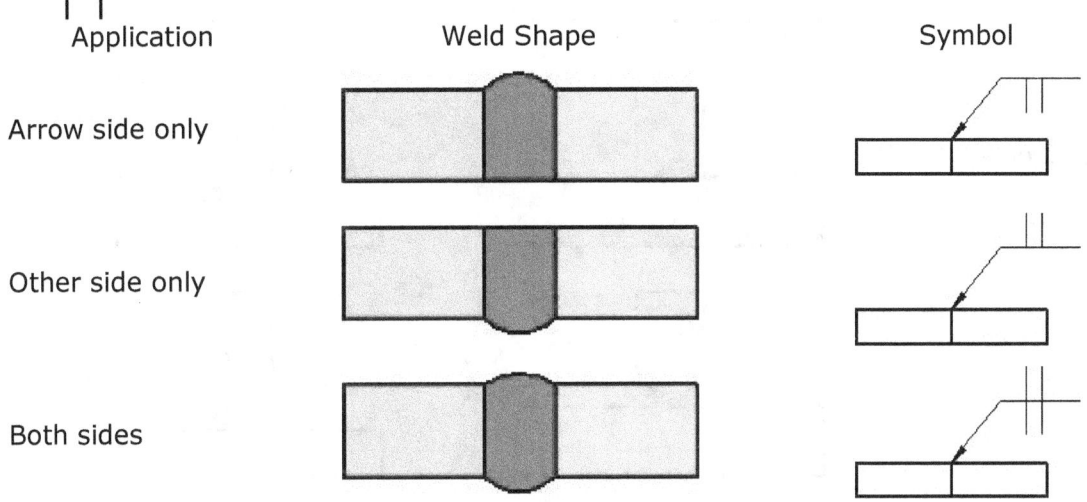

	Weld Shape	Symbol
Arrow side only		
Other side only		
Both sides		

Bevel Butt:

Bevel butt welds are welds where one piece in the joint is beveled at approximately 30° and the other surface is perpendicular to the plane of the surface. These types of joints are used where adequate penetration cannot be achieved with a square-groove and the metals are to be welded in the horizontal position. Double-bevel butt welds are common in arc and gas welding processes. In this type both sides of one of the edges in the joint are beveled. Bevel butt welds allow for greater penetration and can be used on plates measuring 5mm to 15mm thick.

The weld symbol for a bevel butt weld consists of a 5mm long vertical line and a line at 30° to the right.

Butt Welds

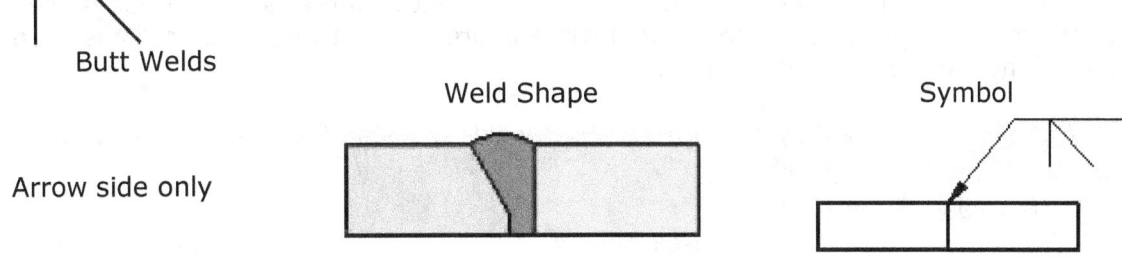

	Weld Shape	Symbol
Arrow side only		

Other side only

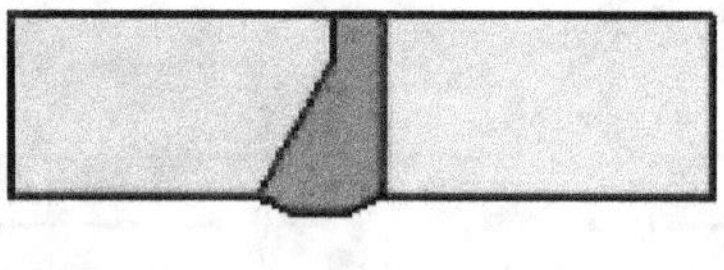

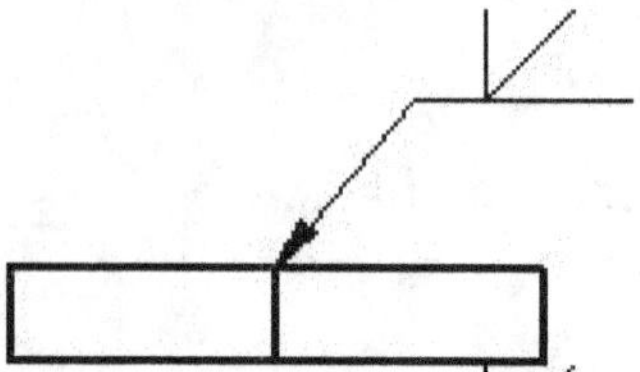

Both sides

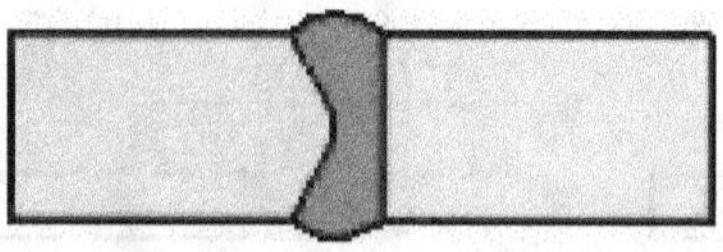

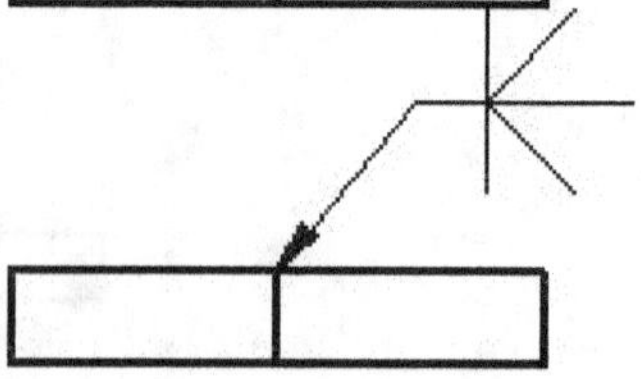

T Joints

Weld Shape   Symbol

Arrow side only

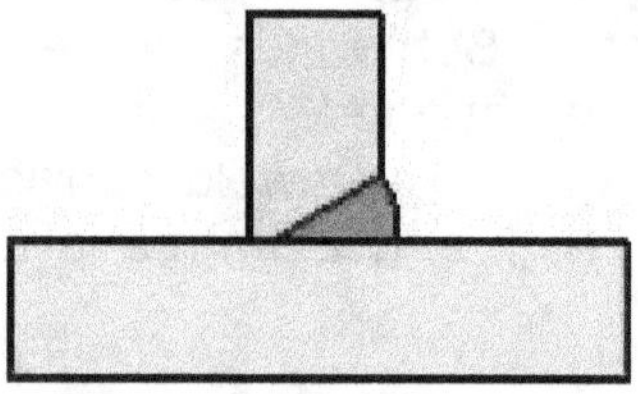

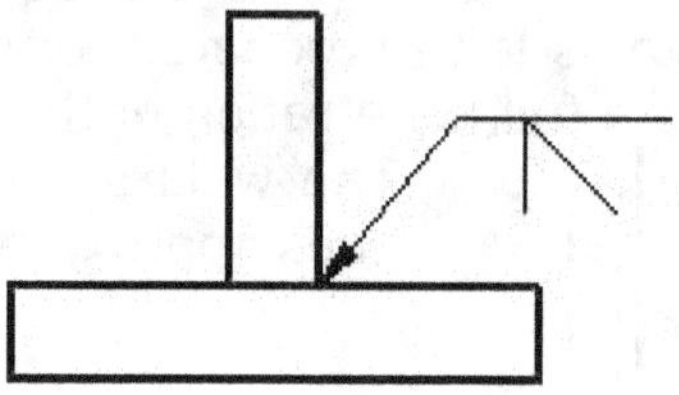

Other side only

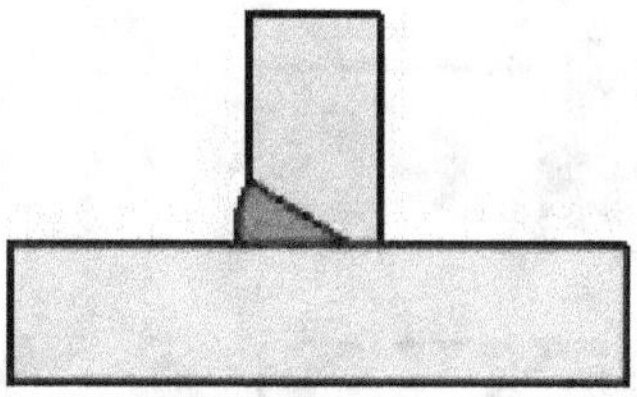

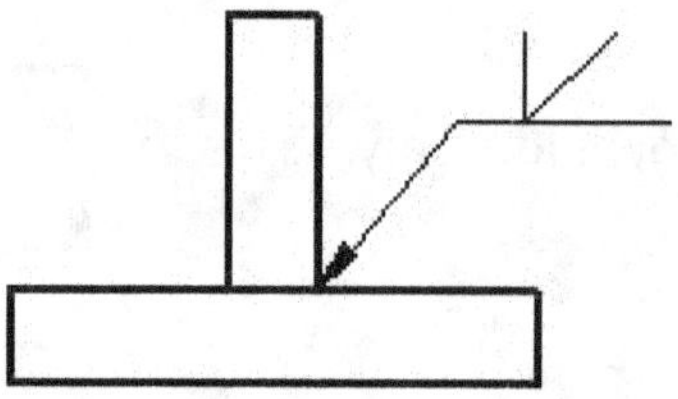

Both sides

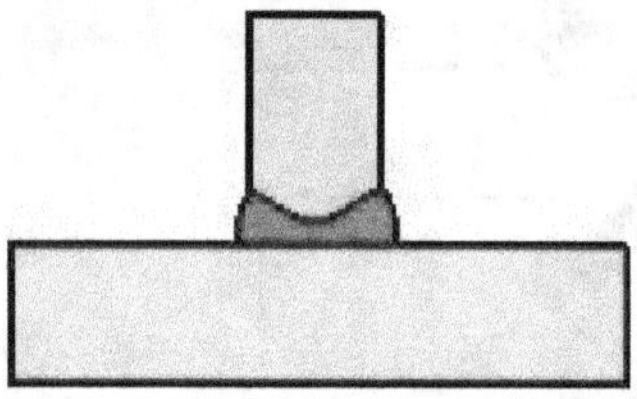

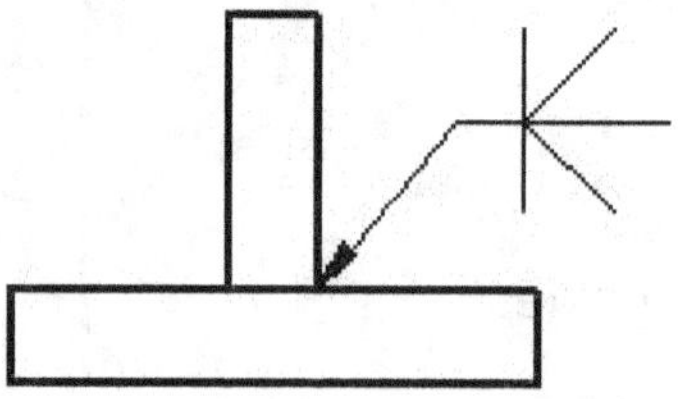

"V" Butt:

The V butt weld is used on plates ranging from 5mm to 20mm thick for single bevels and double V butt welds over 20mm to achieve greater penetration; each member is bevelled so the included angle for the joint is approximately 60 degrees for plate and 75 degrees for pipe. Preparation of the joint requires a special bevelling machine (or cutting torch), which makes it more costly than a square butt joint. It also requires more filler material than the square joint; however, the joint is stronger than the square butt joint. As with the square joint, it is not recommended when subjected to bending at the root of the weld. The double-V butt joint (the plates are bevelled on the both sides of the plate) is an excellent joint for all load conditions. Its primary use is on metals thicker than 20mm but can be used on thinner plate where strength is critical. Compared to the single V joint, preparation time is greater, but less filler metal is used because of the narrower included angle.

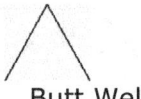

Butt Welds

The weld symbol for a V butt weld consists of 2 – 5mm lines at 60°, to each other.

Weld Shape   Symbol

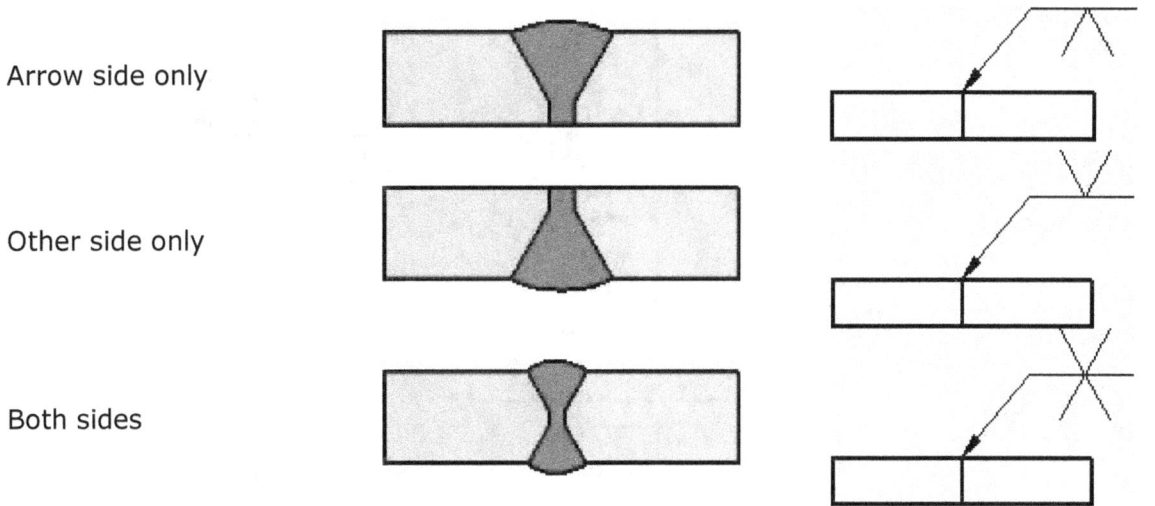

Arrow side only		
Other side only		
Both sides		

"J" Butt:

J butt welds are when one edge of a plate is prepared in the shape of a "J" that easily accepts filler material while the other piece is square. A J-groove is formed either with special cutting machinery or by grinding the joint edge into the form of a J. Although a J-groove is more difficult and costly to prepare than a V-groove, a single J-groove on metal between a 12mm and 20mm thick provides a stronger weld that requires less filler material. Double-J-butt welds have one piece that has a "J" shape on both sides and the other piece is square; double J butt welds are used on materials in excess of 20mm.

The weld symbol for a J butt weld consists of vertical line 5mm long and a ¼ arc, 3mm radius, on the right side of the vertical line. The arc is located 2mm from the reference line

Butt Welds

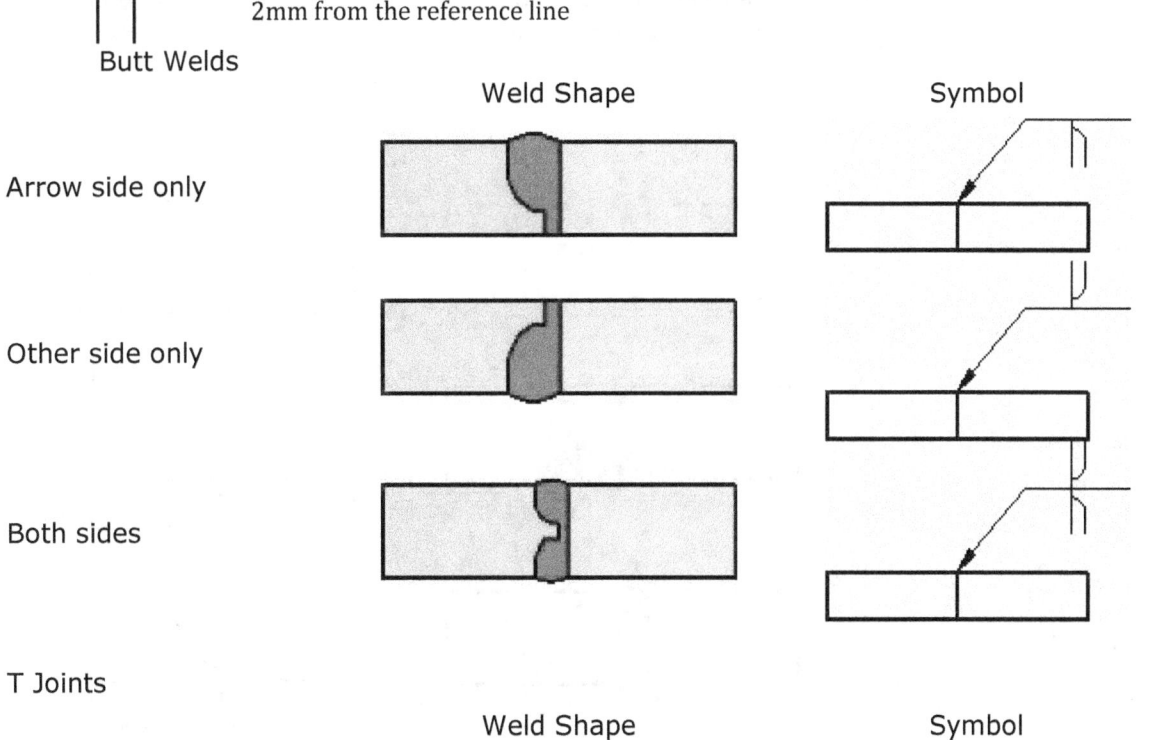

	Weld Shape	Symbol
Arrow side only		
Other side only		
Both sides		

T Joints

	Weld Shape	Symbol

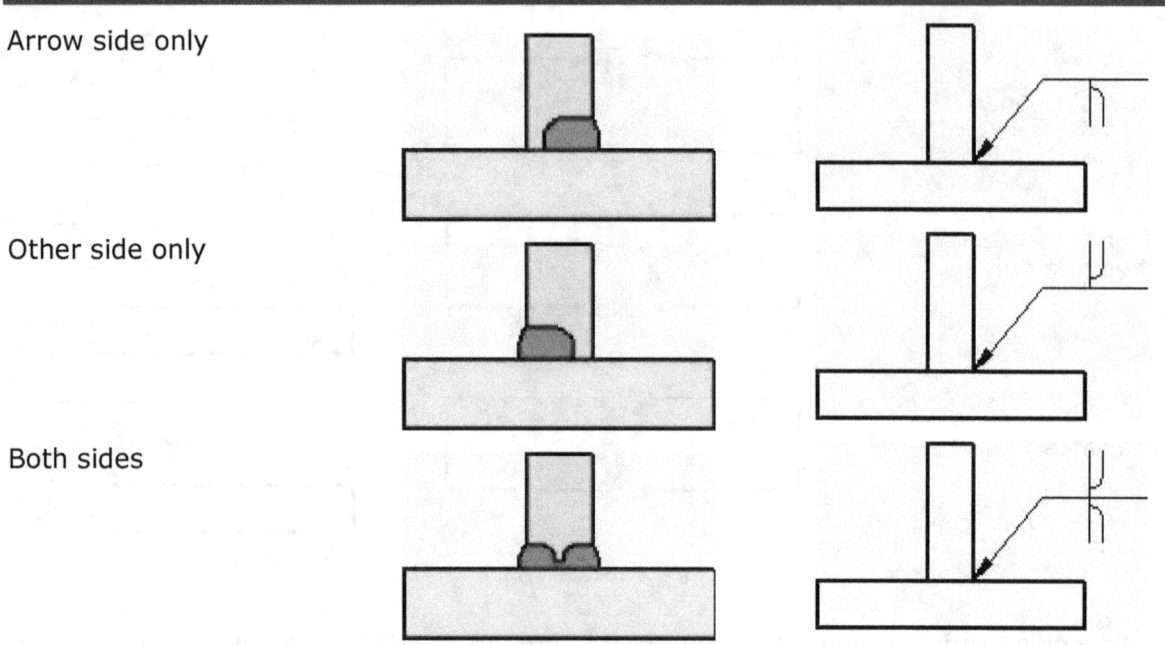

Arrow side only

Other side only

Both sides

"U" Butt:

Single U butt welds are welds that have both edges of the weld surface shaped like a J, but once they come together, they form a U shape. Double-U-Joints have a U formation on both the top and bottom of the prepared plates. U-joints are the most expensive edge to prepare and weld and are usually used on thick base metals where a V-groove would be at such a extreme angle, that it would cost too much to fill. Single U butt welds are used on plates less than 20mm and double U butts on plates over 20mm thick.

The weld symbol for a U butt weld consists of a mirror image of the J butt using the same proportions.

Butt Welds

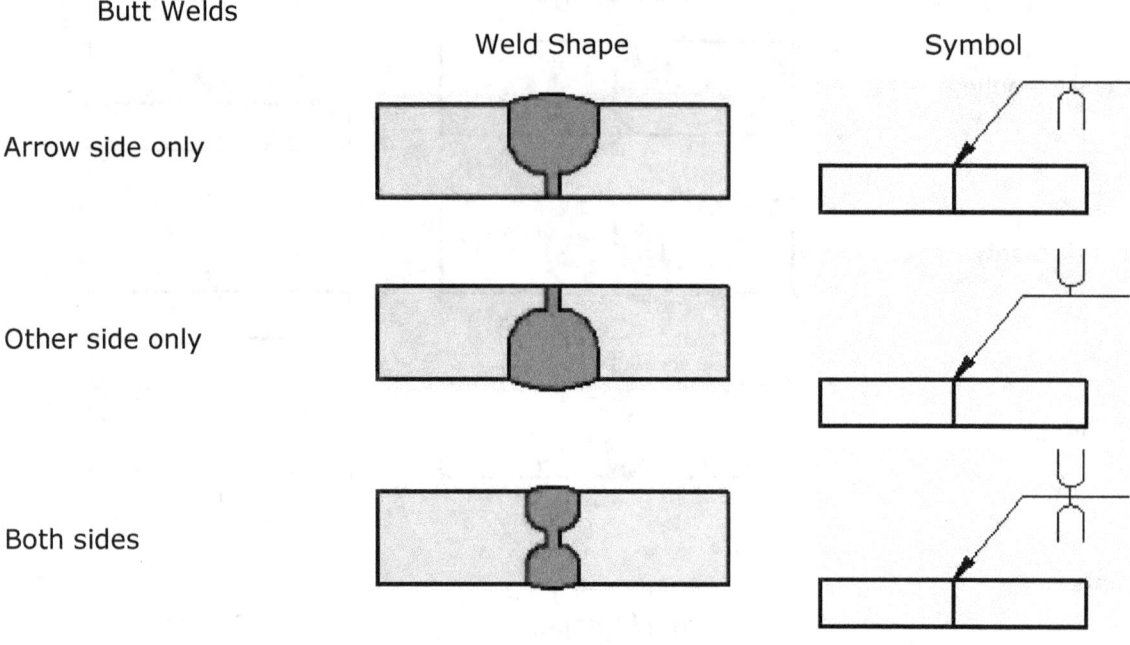

Weld Shape   Symbol

Arrow side only

Other side only

Both sides

Supplementary Welding Symbols:

Supplementary welding symbols convey additional information relative to the extent of the welding, where the welding is to be performed, and the contour of the weld bead.

All Around Weld:

The All-Around symbol is a circle at the intersection of the leader and reference liness and indicates the weld is "ALL AROUND"; this means the weld extends all the way around the joint to which the arrow is pointing. The all around symbol is only used when it is possible to weld all the way around a single surface otherwise more than on symbol is used.

The weld symbol for an all-around weld consists of an Ø3 circle.

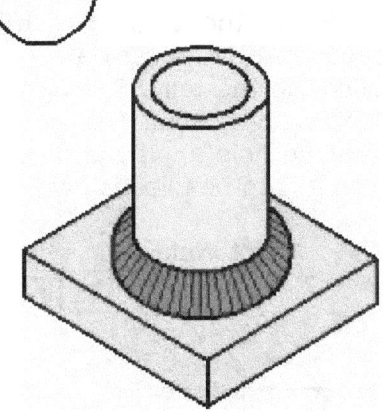
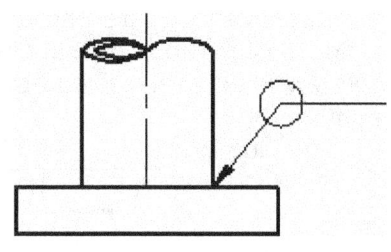

On-Site Weld:

Welds that cannot be made in the shop but must be welded at the site location of the project are identified as Field or On-Site welds and appear as flags placed at the intersection of the leader and reference lines. On older drawings the symbol may appear as a solid dot at the intersection of the leader and reference lines.

The weld symbol consists of a solid filled 3mm triangle at the end of a vertical 6mm line; the triangle always points to the left.

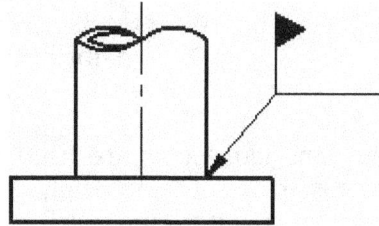

Stud Weld:

Stud welding is a process in which an arc is struck between the bottom of a stud and the base metal. When a pool of molten metal has formed, the arc is extinguished and the stud is driven into the pool to form a weld.

The weld symbol for a stud weld consists of a 3mm long horizontal and a 5mm high vertical line; the horizontal line is placed 1mm to 2mm from the reference line.

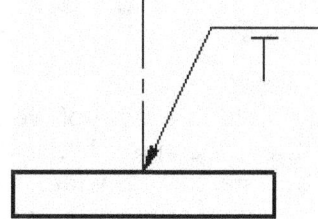

Plug & Slot Welds:

Plug welds and slot welds are used join overlapping members, one of which has holes (round for plug welds, elongated for slot welds) in it. Weld metal is deposited in the holes and penetrates and fuses with the base metal of the two members to form the joint. Plug and slot welds are used where a person or automatic welding machine cannot gain access to a space to apply a weld.

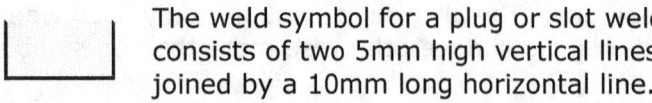 The weld symbol for a plug or slot weld consists of two 5mm high vertical lines joined by a 10mm long horizontal line.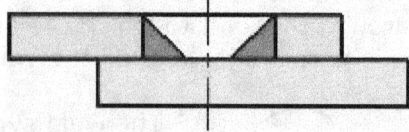

For plug a weld, the diameter of each plug is given to the left of the symbol and the plug-to-plug spacing (pitch) is given to the right. For slot welds, the width of each slot is given to the left of the symbol, the length and pitch (separated by a dash) are given to the right of the symbol, and a detail drawing is referenced in the tail. The number of plugs or slots is given in parentheses above or below the weld symbol. The arrow-side and other-side designations indicate which piece contains the hole(s). If the hole is not to be completely filled with weld metal, the depth to which it is to be filled is given within the weld symbol.

Plug Weld   Slot Weld

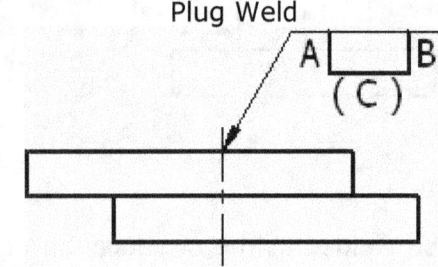   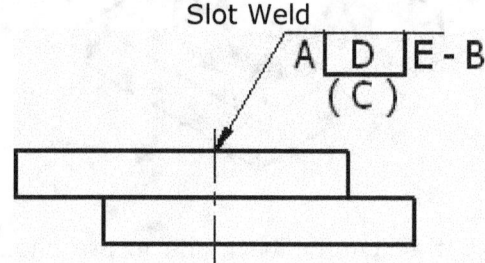

Where:

    A = Diameter or width of the slot.
    B = Pitch of the plug or slot.
    C = Number of plugs or slots.
    D = Height of the weld.
    E = Length of the slot.

Plug and slot welds are used where a person or automatic welding machine cannot gain access to a space to apply a weld. In Figure 19 below, the outer plates could be the lining of a cofferdam which must be welded to the beams. Plug and slot welds would be required as shown because the welds indicated at X cannot be accessed.

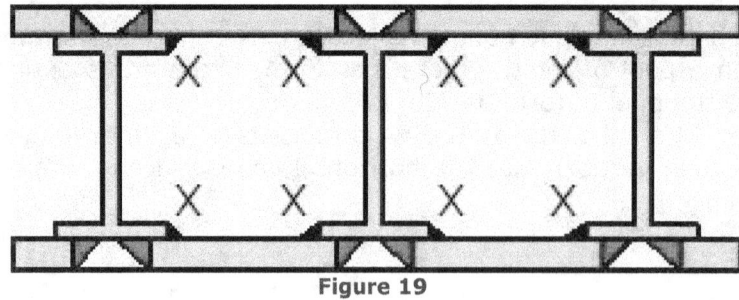

Figure 19

Bead Weld:

A bead weld is a type of weld composed of a single string of weld deposited on a plate, or on the underside of a weld to clean up the appearance and surface. A weld bead may be either narrow or wide, depending on the amount of transverse oscillation (side-to-side movement) used by the welder. When there is a great deal of oscillation, the bead becomes wide; when there is little or no oscillation, the bead is narrow.

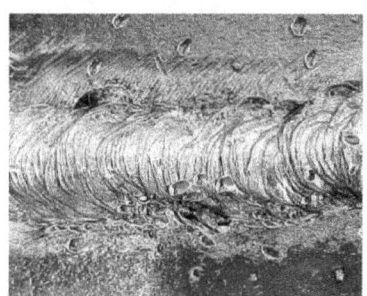

The weld symbol for a bead weld consists of an R1.5 arc placed on the reference line.

The image to the right shows a typical bead on a surface.

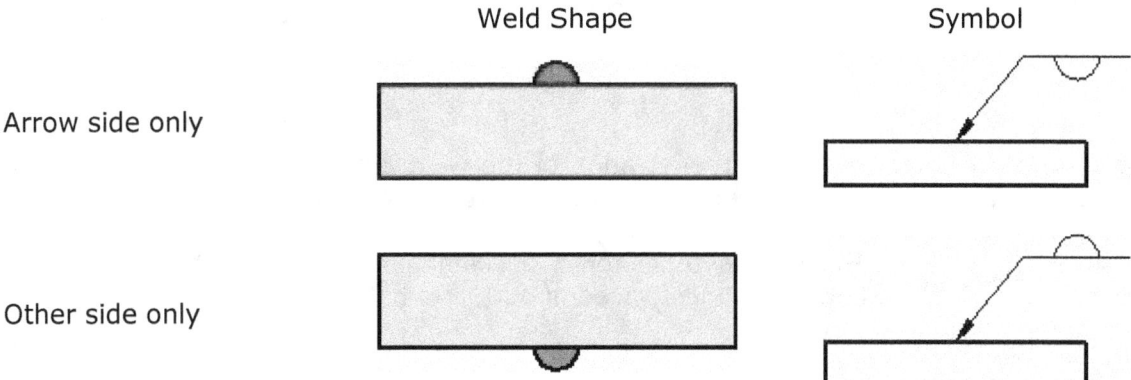

	Weld Shape	Symbol
Arrow side only		
Other side only		

A bead weld is mainly used in conjunction with a butt weld to clean up the underside of the weld. The following examples indicate a bead applied to a bevel butt weld.

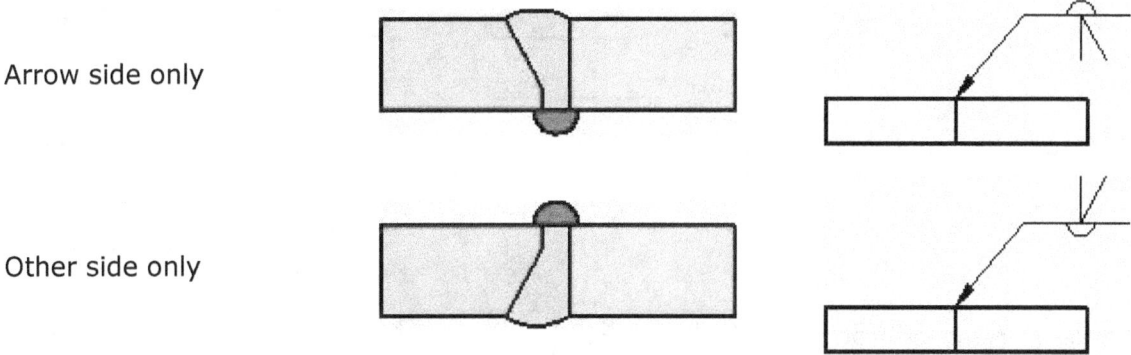

Arrow side only		
Other side only		

Surfacing:

Surfacing is a welding process used to apply a hard, wear-resistant layer of metal to surfaces or edges of worn-out parts. It is one of the most economical methods of conserving and extending the life of machines, tools, and construction equipment. Surfacing, sometimes known as hard facing or wear facing, is often used to build up worn shafts, gears, cutting edges or pitted areas of steel surfaces.

The weld symbol for a bead weld consists of three R1.5 arcs placed on the reference line.

The image to the right shows surfacing on a surface.

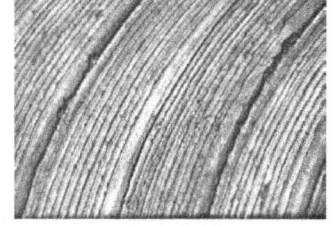

Weld Shape   Symbol

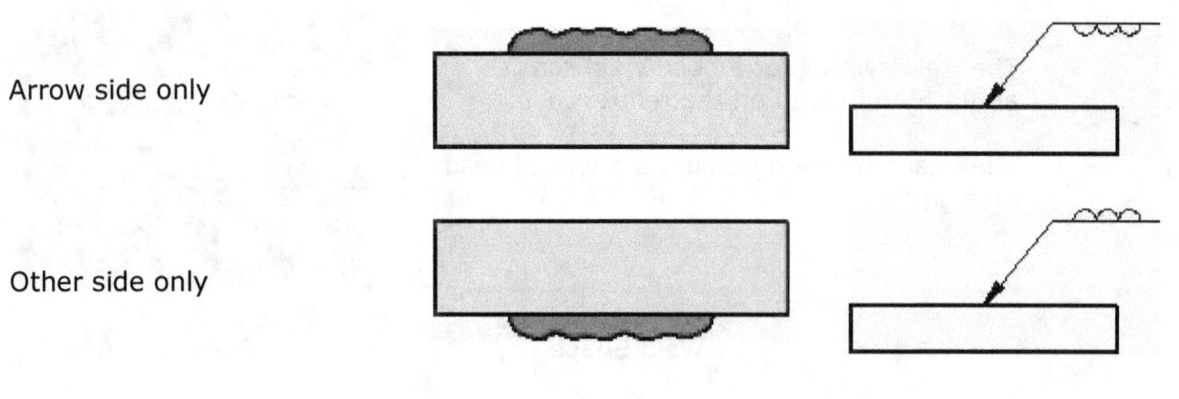

Arrow side only

Other side only

Backing Bar:

A backing bar is used to gain full penetration of the weld, ensuring a strong weld. The backing bar can be left permanently on the structure, or removed after the welding process.

———— Permanent   The weld symbol for a backing bar consists of two 5mm long horizontal lines placed about 2mm below the reference line.

‒ ‒ ‒ Removed

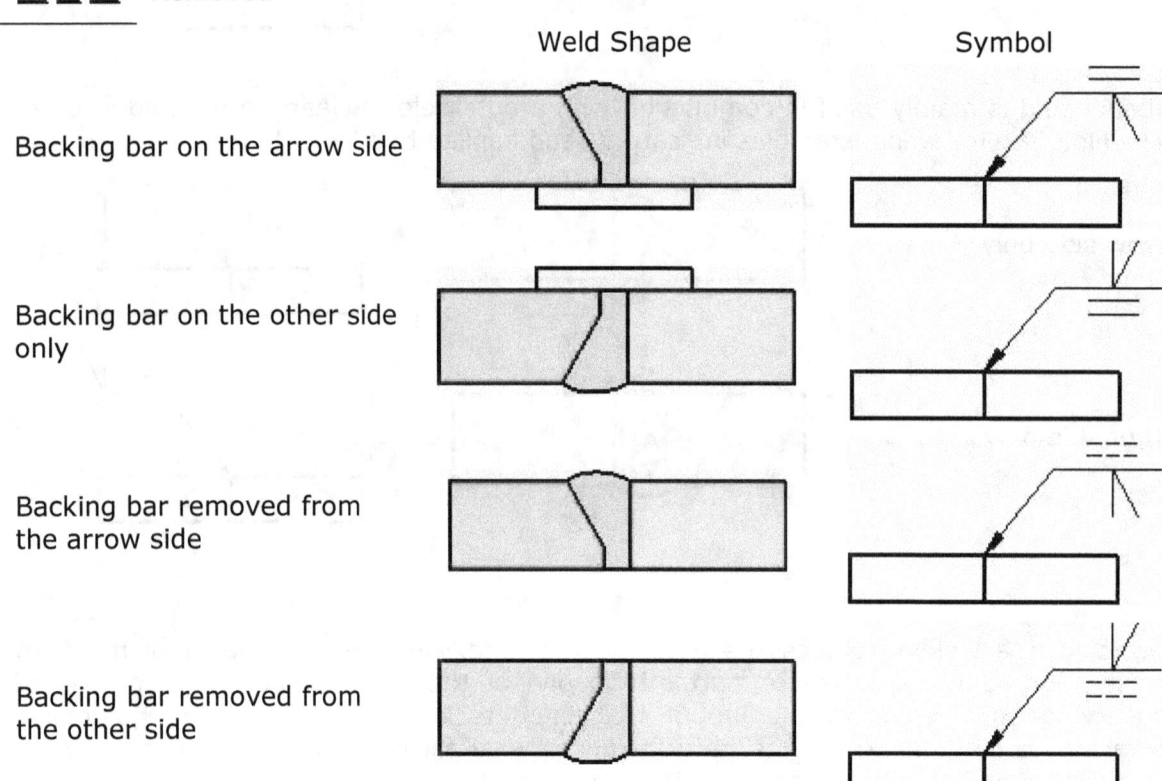

	Weld Shape	Symbol
Backing bar on the arrow side		
Backing bar on the other side only		
Backing bar removed from the arrow side		
Backing bar removed from the other side		

If the backing bar is to be left permanently attached to the plate, it should be held in place by fillet welds.

Flush, Convex & Concave Surface Contour:

At times the finished surface of a weld may be required to be subjected to a machining process. Most welds leave a convex surface and normally need no finishing apart from scrubbing with a wire brush to remove any loose particles. If the joint is in a rail, engine bed or floor plate, the finished weld surface may need to be level with the parent material to ensure smooth movement of wheels or goods. A weld may occasionally be required ground to below the parent material with a small curve.

_____	Flush	A horizontal line located just off the weld symbol
⌢	Convex	An arc located just off the weld symbol; as the finish is natural, the symbol is mostly omitted unless required.
⌣	Concave	An arc located just off the weld symbol

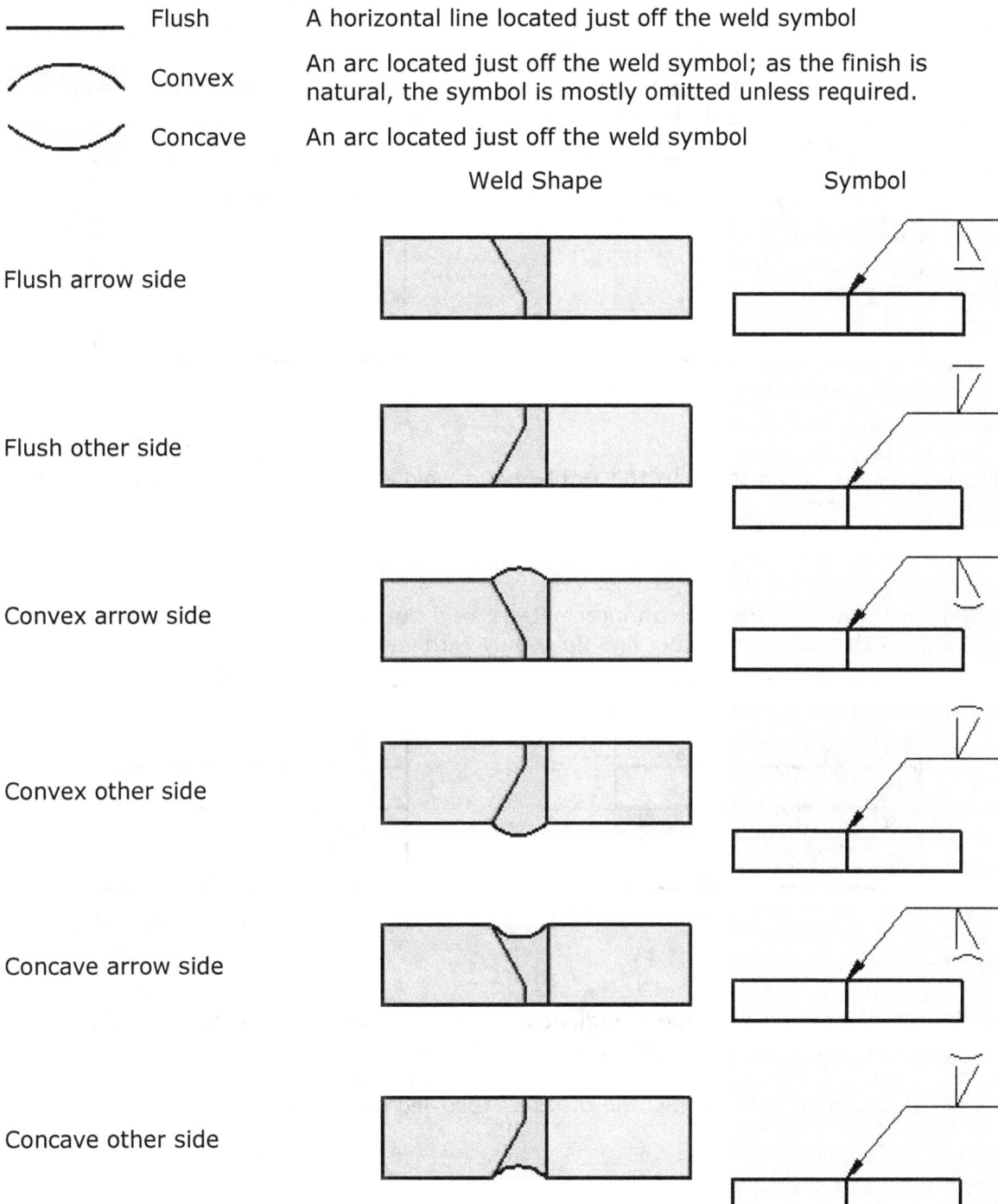

	Weld Shape	Symbol
Flush arrow side		
Flush other side		
Convex arrow side		
Convex other side		
Concave arrow side		
Concave other side		

Tail:

The tail of the weld symbol is the place for supplementary information on the weld. It may contain a reference to the welding process, the electrode, a detail drawing, any information that aids in the making of the weld that does not have its own special place on the symbol.

Note

Ancillary Welding Symbols:

Ancillary welding symbols include intermittent, staggered and compound welds.

Intermittent Weld:

An intermittent weld is one that does not require a continuous weld along the length of the entire joint. Intermittent welds are shorter welds, spaced evenly along the joint.

Weld Shape	Symbol
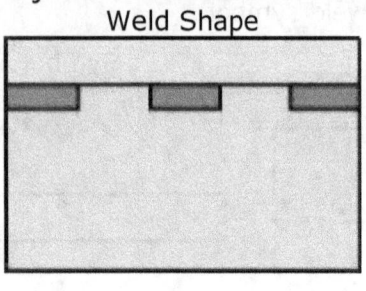	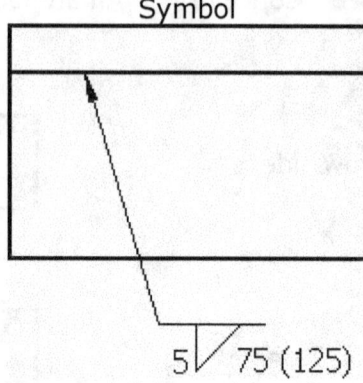

The weld length is indicated to the right of the weld symbol followed by the unwelded length with the brackets.

Staggered Weld:

A staggered weld is similar to an intermittent weld but requires a sequence of welds on each side of the joint that do not line up evenly with each other.

Weld Shape	Symbol
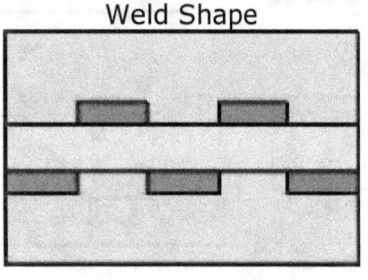	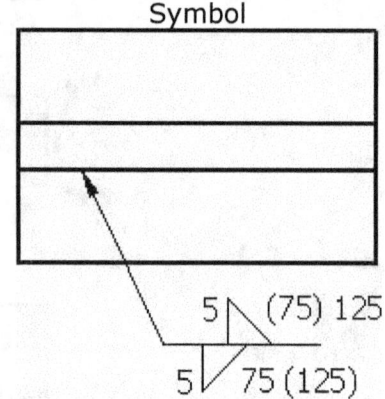

The length of weld on one side is indicated to the right of the weld symbol followed by the unwelded length with the brackets as per an intermittent weld. The symbol on the opposite side of the reference line is moved slightly to show the stagger with the unwelded length indicated within the brackets followed by the length of weld.

Compound Weld:

Compound welds are used where a greater penetration is required and gives a stronger weld; they are generally used in T-Joints where a butt weld is first applied followed by a fillet weld.

Arrow side

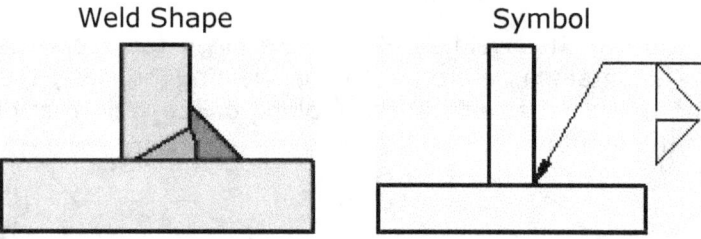

Weld Shape   Symbol

Other side

Skill Practice Exercise   MSATCS302-SP-301 to MSATCS302-SP-303:

MSATCS302-SP-301:

Prepare a detail drawing of the Wall Stand on an A3 sheet including the nominated welds.

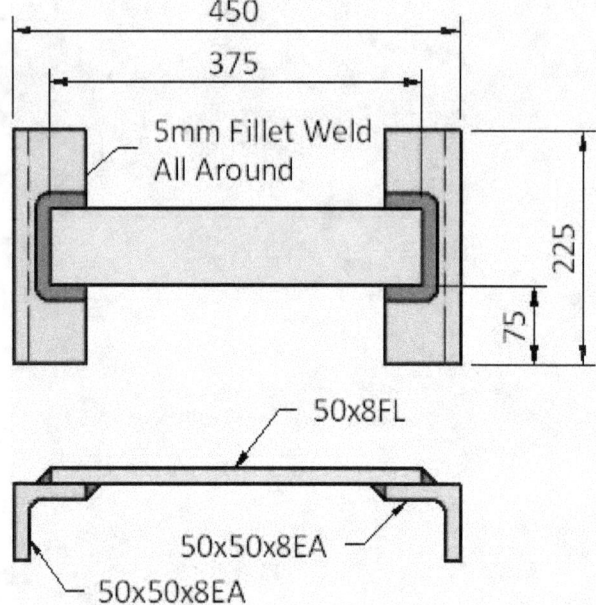

MSATCS302-SP-302:

Prepare a detail drawing of the Hopper Gate on an A3 sheet including the nominated welds.

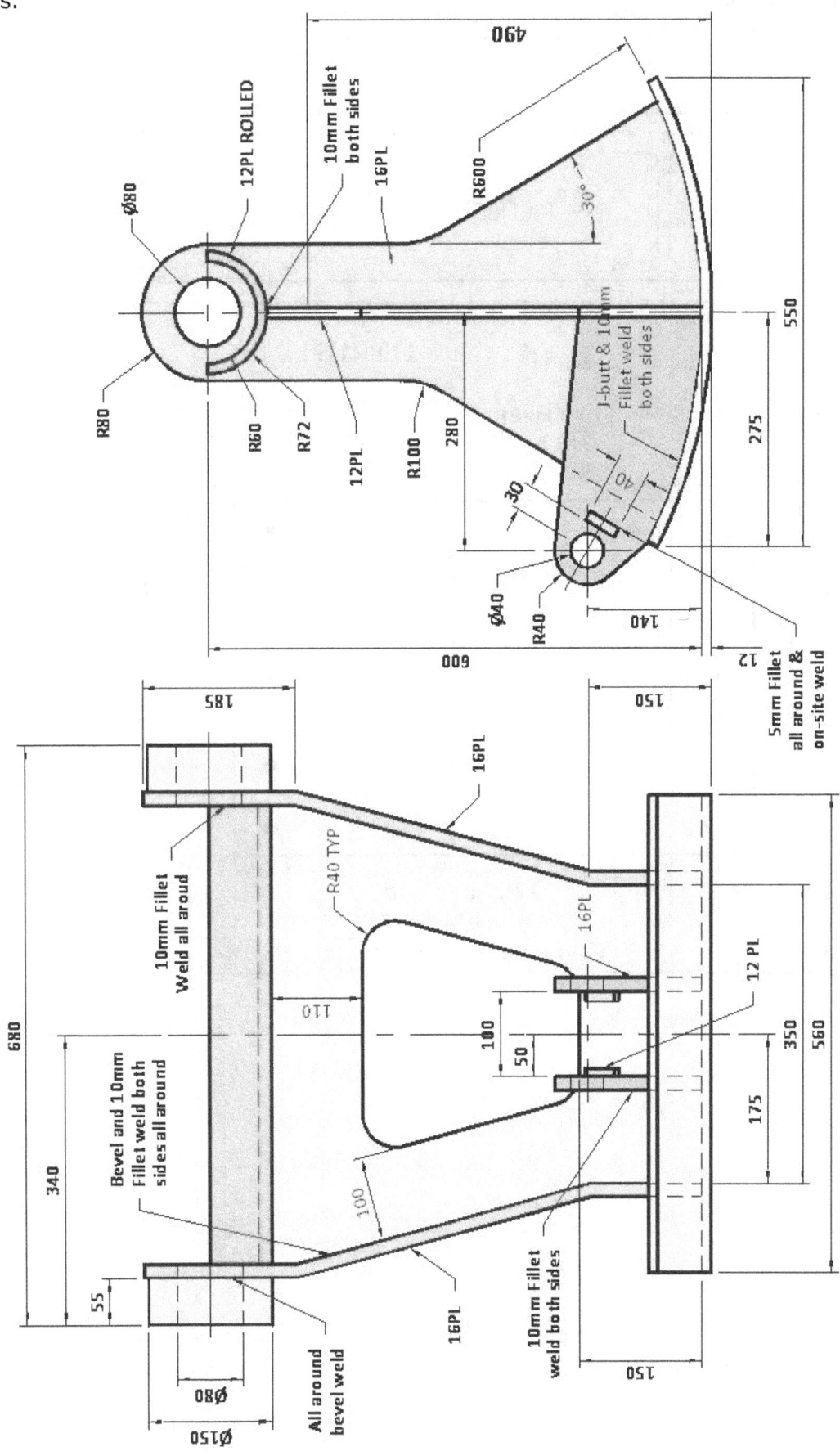

MSATCS302-SP-303:

Prepare a detail drawing of the Advertising Hording on an A3 sheet including the nominated welds.

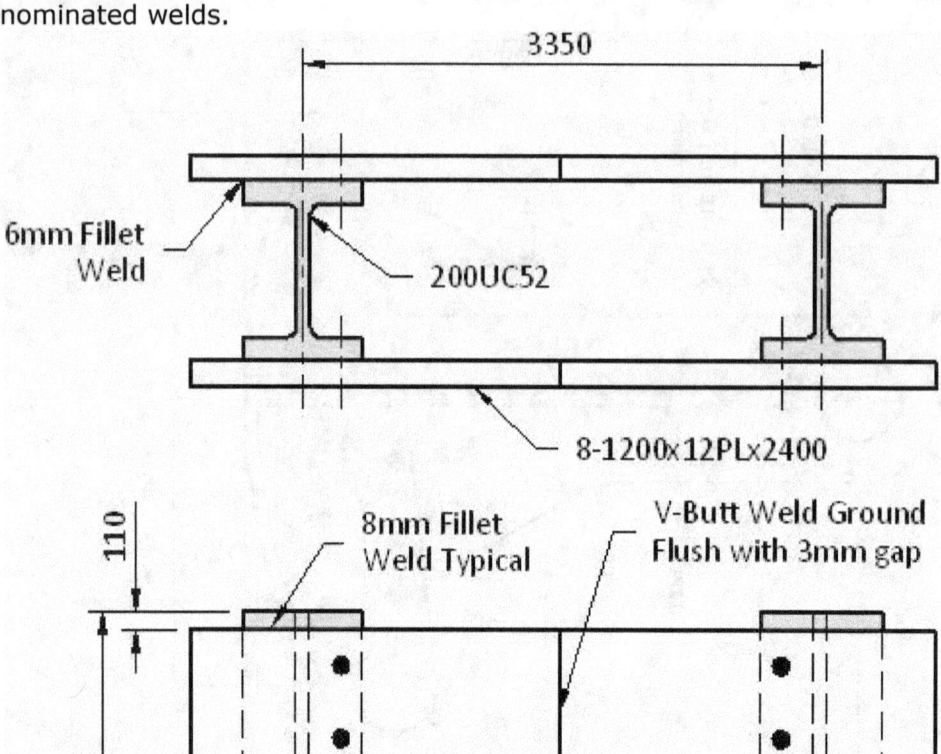

Page Left Deliberately Blank

Page Left Deliberately Blank

Practice Competency Test:

- Reproduce the drawing of the Filter Frame shown below on a standard A3 sheet indicating all welds, holes and dimensions where necessary.
- Create a detail drawing of the 150x16x75 Unequal Angle.
- Call up the bolts to be used on the assembly where the bolts are spaced at their minimum spacing allowed. The length of the Filter Frame is 2400mm long.
- Show a Front View and Sectional View at different scales. Drawing number is MSATCS302-PT-01

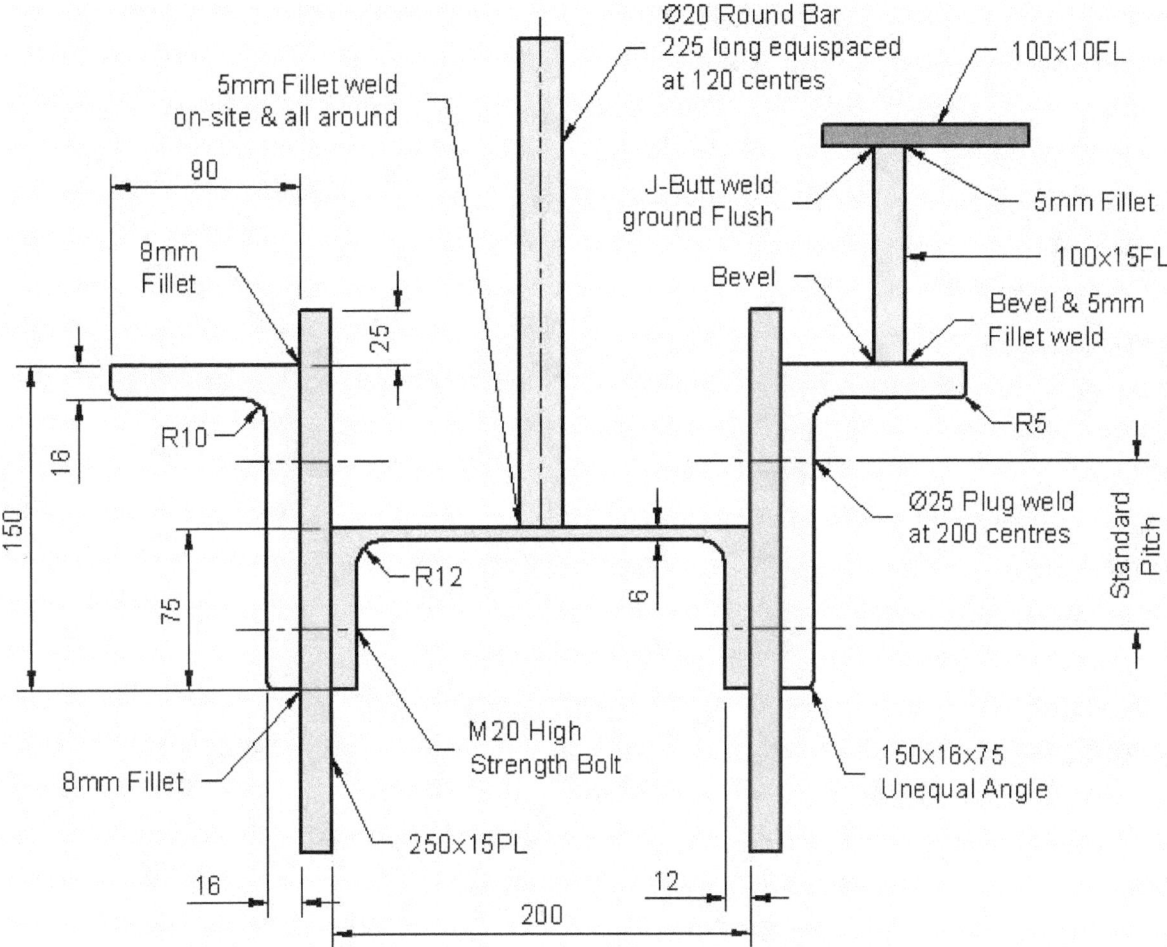

A student will be assessed as competent if the drawing has been marked-up by the teacher and corrections completed.

Page Left Deliberately Blank

Page Left Deliberately Blank

Addendums:

Addendum 1 – Metric Hexagon Commercial Bolts & Set Screws:

Thread ISO Metric Coarse Pitch Series
Thread Class 8g, Property Class 4.6
Dimensions to AS1111-1996

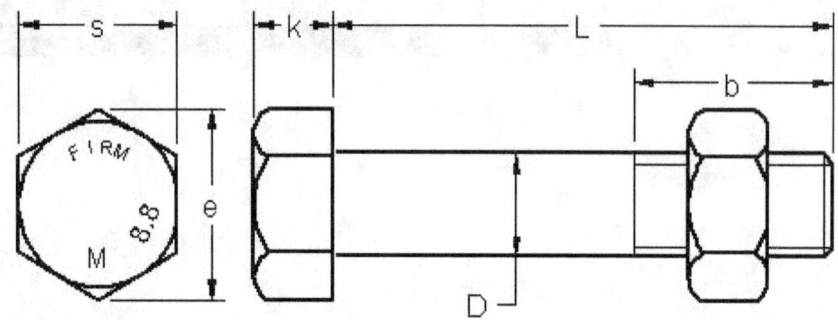

Size D	Pitch of Thread	Body Diameter (On Bolts) Ds		Width Across Flats s		Head Thickness k		Across Corners e
		Max	Min	Max	Min	Max	Min	Min
M6	1.0	6.48	5.52	10.0	9.64	4.38	3.62	10.89
M8	1.25	8.58	7.42	13.0	12.57	5.68	4.92	14.20
M10	1.5	10.58	9.42	16.0	15.57	6.85	5.95	17.59
M12	1.75	12.70	11.30	18.0	17.57	7.95	7.05	19.85
M14	2.0	14.70	13.30	21.0	20.16	9.25	8.35	22.78
M16	2.0	16.70	15.30	24.0	23.16	10.75	9.25	26.17
M18	2.5	18.70	17.30	27.0	26.16	12.40	10.60	29.56
M20	2.5	20.84	19.13	30.0	29.16	13.40	11.60	32.95
M22	2.5	22.84	21.16	34.0	33.00	14.90	13.10	37.29
M24	3.0	24.84	23.16	36.0	35.00	15.90	14.10	39.55
M27	3.0	27.84	26.16	41.0	40.00	17.90	16.10	45.20
M30	3.5	30.84	29.16	46.0	45.00	19.75	17.65	50.85
M33	3.5	34.00	32.00	50.0	49.00	22.50	19.95	55.37
M36	4.0	37.00	35.00	55.0	53.80	23.55	21.45	60.79
M36	4.0	40.00	38.00	60.0	58.80	26.05	23.95	66.44
M42	4.5	43.00	41.00	65.0	63.10	27.67	24.35	71.30
M48	5.0	49.00	47.00	75.0	73.10	31.65	28.35	82.60
M56	5.5	57.20	54.80	85.0	82.80	36.95	33.05	93.56
M64	6.0	65.20	62.80	95.0	92.8	41.95	39.05	104.86

All dimensions are in millimetres. See Addendum 5 – Minimum Thread Lengths: for nominal thread lengths.

Addendum 2 – High Strength Structural Bolts:

Property Class 8.8
Thread ISO Metric Coarse Pitch
Series Dimension to AS1252-1996

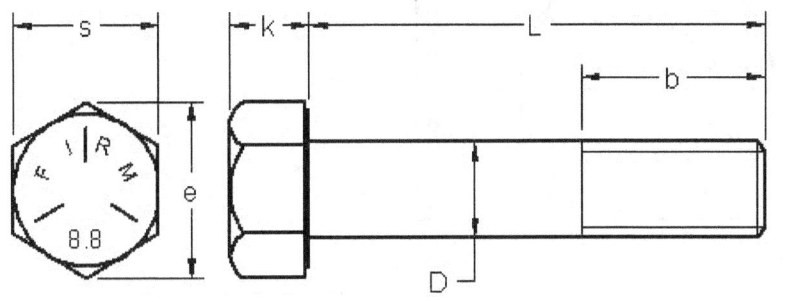

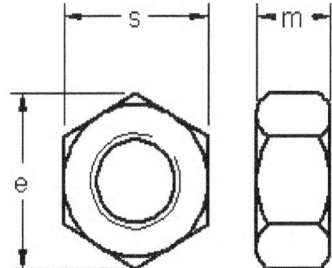

Bolt

Nut

Size	Pitch of Thread	Body Diameter		Width Across Flats		Width Across Corners	Head Thickness		Width Across Flats		Thickness	
Bolt Dimensions									**Nut Dimensions**			
D		D1		s		e	k		s		m	
		Max	Min	Max	Min	Min	Max	Min	Max	Min	Max	Min
M16	2.0	16.70	15.3	27	26.16	29.59	10.75	9.25	27	26.16	17.1	16.0
M20	2.5	20.84	19.16	32	31.00	35.03	13.90	12.10	32	31.00	21.3	20.0
M24	3.0	24.84	23.16	41	40.00	45.20	15.90	14.10	41	40.00	25.3	24.0
M30	3.5	30.84	29.16	50	49.00	55.37	19.75	17.65	50	49.00	31.3	30.0
M36	4.0	37.00	35.00	60	58.80	66.44	23.55	21.45	60	58.80	37.6	36.0

All dimensions are in millimetres. See Addendum 5 – Minimum Thread Lengths: for nominal thread lengths. For details refer to AS1252-1983

Addendum 3 – Coronet Load Indicators

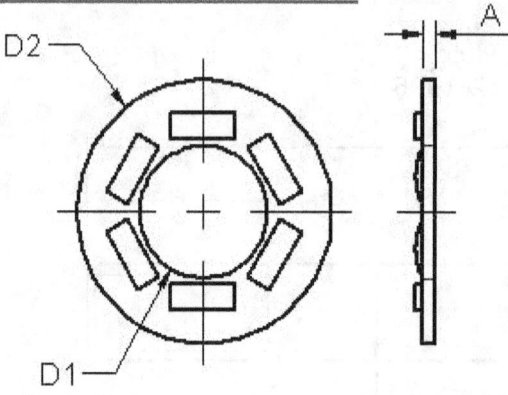

Nominal Bolt Diameter	Outside Diameter D2	Inside Diameter D1	Thickness A (max)	Minimum Bolt Tension kN
M16	35.45	13.70	4.26	100
M20	41.67	20.84	4.26	150
M24	50.69	24.84	4.25	220
M30	59.59	30.84	4.26	350
M36	80.00	37.50	4.26	515

All dimensions are in millimetres

Note: A nut should not be slackened after fully tightening with a Load Indicator. If any slackening is necessary, a new Load Indicator is to be used for the second tightening.

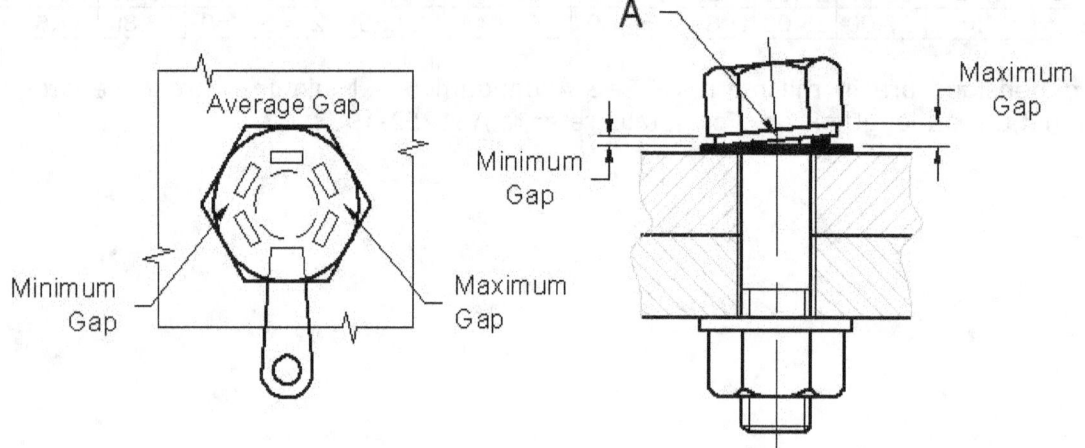

Load Indicator Gaps to Give Required Minimum Shank Tension.

Load Indicator Fitting	AS4100 (1511)
Under bolt head black finish bolts	0.40mm
All plating except galvanised bolts	0.40mm
Galvanised bolts	0.25mm
Under nut with hard flat washer. Black and all flat washer coatings	0.25mm

Addendum 4 – Flat Round Washers for High Strength Bolts:

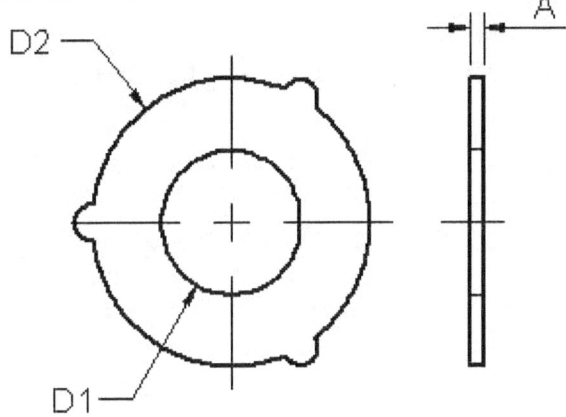

Nominal Bolt Diameter	Inside Diameter D1		Outside Diameter D2		Thickness A	
	Max.	Min.	Max.	Min.	Max.	Min.
M16	18.43	18.0	34.0	32.4	4.6	3.1
M20	22.52	22.0	39.0	37.4	4.6	3.1
M24	26.52	26.0	50.0	48.4	4.6	3.4
M30	33.62	33.0	60.0	58.1	4.6	3.4
M36	39.62	39.0	72.0	70.1	4.6	3.4

All dimensions are in millimetres

Addendum 5 – Minimum Thread Lengths:

Metric Hexagon Commercial Bolts & Set Screws:
Nominal thread lengths for bolts. Set screws are threaded to within 2½ pitches of the head.

Nominal Length of Bolt - L	Minimum Length of Thread - b
Up to and including 125mm	2D + 6mm
Over 125mm up to and including 200mm	2D + 12mm
Over 200mm	2D + 25mm

Where D = Nominal Diameter in millimetres.

High Strength Structural Bolts & Set Screws:
Nominal thread lengths for bolts. Set screws are threaded to the head.

Nominal Length of Bolt - L	Minimum Length of Thread - b
Up to and including 125mm	2D + 6mm
Over 125mm up to and including 200mm	2D + 12mm
Over 200mm	2D + 25mm

Where D = Nominal Diameter in millimetres.

Table 4 – Thread Screw Pitches:

Ø	Inch Series				ISO Metric Preferred Coarse Pitch Series	
Diameter in inches	Threads per Inch				Dia. In mm	Pitch in mm
	BSW	BSF	UNC	UNF		
No. 8	-	-	32	36	1.6	0.35
No. 10	-	-	24	32	2	0.4
3/16	24	32	-	-	2.5	0.45
1/4	20	26	20	28	3	0.5
5/16	18	22	18	24	4	0.7
3/8	16	20	16	24	5	0.8
7/16	14	18	14	20	6	1
1/2	12	16	13	20	8	1.25
9/16	12	16	12	18	10	1.5
5/8	11	14	11	18	12	1.75
3/4	10	12	10	16	16	2
7/8	9	11	9	14	20	2.5
1	8	10	8	12	24	3
1-1/8	7	9	7	12	30	3.5
1-1/4	7	9	7	12	36	4
1-3/8	6	8	6	12	42	4.5
1-1/2	6	8	6	12	48	5
1-5/8	5	8	-	-	56	5.5
					64	6

Page Left Deliberately Blank

Page Left Deliberately Blank

Page Left Deliberately Blank